Seeing the Social

Selected visibility technologies

Harry Freemantle

First published by Vivid Publishing P.O. Box 948, Fremantle Western Australia 6959 www.vividpublishing.com.au

This edition published on Amazon/CreateSpace 2013

National Library of Australia Cataloguing-in-Publication data Author: Freemantle, Harry. Title: Seeing the social: selected visibility technologies / Harry Freemantle. Edition: 1st ed. ISBN: 9781921787164 (pbk.) Subjects: Science--Social aspects. Technology--Social aspects. Vision--Technological innovations. Dewey Number: 303.483

CONTENTS

Introduction

Bibliography

INTRODUCTION

In *Seeing the Social* I look closely at the early moments of visibility technologies like perspective, lenses (especially microscopes), the camera obscura, The *Encyclopédie*, the balloon, the lithograph, the diorama and photography, including the accompanying metaphors, in order to draw out how the visual aspects of seeing inform the articulable at particular moments in history. How these visibilities shift and change over time can then be charted against larger social movements. Such visibility technologies formed part of the epistemological conditions of the observer, underlay the discourse and contributed to how early writers on the social saw the emerging social world.

Seeing the Social is historically rigorous and where possible relies on primary texts. What is different about what it does with these moments is the attempt to tie these technologies to seeing the social and allowing the narrative, sometimes in dialogue form, to convey the story. I move away from the idea that grand philosophical or theoretical knowledge accounts tell the complete story, preferring rather to allow some minor actors to shed light on the subject. The book also attempts to steer away from being theory or theorist driven. Rather it allows these early moments of surprise and wonderment to be heard. To this end I do not labour my point nor overdraw my analyses, preferring to allow the visual spaces to emerge for the reader in an effort to encourage further research and reading.

Each of the visibility technologies are presented in a roughly chronological way, starting with perspective and ending with photography. Similarly the content of each chapter begins with the early days of invention and moves towards later discoveries and uses. Having said this some key names and technologies spill between chapters; they are not mutually exclusive, and I do not always hold fast to chronology within the chapters, for example during the chapter on photography some early letters appear towards the end.

The book begins with a discussion I have with the Italian poet, scholar, architect, art theorist, engineer and mathematician Leon Battista Alberti on his recent publication *Della Pictura*. Alberti claims that the sole purpose of his 1436 text is to establish the rudiments of optical geometry external to the eye and the consequences this has for the painter. Despite being well read in medieval optical science, including the work of people like the Islamic philosopher Ibn-Al-Haithan, the Franciscans Roger Bacon and John Pecham, and Witelo from Poland, he contains himself to a visual description of the material world and obliquely refer to 'those philosophers who are experts.' For Alberti the science provided by these people offers a perfect illustration of the way in which God's rational design insinuates itself throughout nature. The main focus in this chapter is on perspective.

Chapter Two examines lenses. It is not a chronological history of the use of lenses but rather a selection of texts and illustrations that help describe how the microscopic, and to a lesser degree the macroscopic lenses, make possible a vision of the social. This is done by describing the experience of entering a vast hall of luminary and amateur and weekend scientists and others who are fascinated by the natural world and the possibilities opening up to seeing the minute and infinite (and without the aid of a lens an invisible) world they inhabit. What I do is wander around listening to snatches of conversations, joining some that I find useful and look at the various displays and publications on show.

Chapter Three begins with Daniel Barbaro's description of a camera obscura fitted with a lens and the first description of an adjustable diaphragm to sharpen the image. His treatise on perspective was extremely influential during the sixteenth century and the improvement in image definition brought about by the inclusion of a lens and adjustable diaphragm made it useful for tracing. The middle section is taken up by Kepler's posthumously published *Somnium* where he presents a heliocentric world-picture, based on the observations of Copernicus, and inverts the meaning and place of observation in the hierarchy of knowledge to the top. Kepler manages this by critically adapting Lucian's *True Story* and Plutarch's *The Face on the Moon* along with the help of Maurolico's optics. Astronomical knowledge is justified while scientific observation offers a way out of his dream.

Chapter Four opens with the Frontpiece to the 1751 edition of the *Encyclopédie*, drawn by Charles-Nicolas Cochin and engraved by Bonaventure-Louis Prévost. The accompanying caption explains the illustration;

Beneath a temple of Ionic architecture, sanctuary of Truth, we see Truth wrapped in a veil, radiant with a light which parts the clouds and disperses them. On the right of Truth, Reason and Philosophy are engaged, the one lifting the veil from Truth, the other in pulling it away. At her feet Theology, on her knees, receives her light from on high. Following the line of figures, we see grouped on the same side Memory, and Ancient and Modern History; History is writing the annals, and Time serves as a support for her. Grouped below are Geometry, Astronomy, and Physics. The figures below this group show Optics, Botany, Chemistry, and Agriculture. At the bottom are several Arts and Professions that proceed from the Sciences. On the left of Truth we see Imagination, who is preparing to adorn and crown Truth. Beneath Imagination, the Artist has placed the different genres of Poetry – Epic, Dramatic, Satiric, and Pastoral. Next come the other Arts of Imitation – Music, Painting, Sculpture, and Architecture.

These two opening elements set the tone for the *Encyclopédie* project that was to last until 1780 by which time the prime mover and editor, Diderot, and the author of the *Preliminary Discourse* and the sections on mathematics and physics, amongst other things, d'Alembert, were no longer involved. The *Encyclopédie* is a visibility technology of the highest order.

Chapter Five on Balloons broadens the horizon of these technologies by supplying an important spacial component, forming the ground on which the mapping of the world is set. The balloon offers the ultimate in panoramic vision. It has the ability to rise up and over natural barriers such as mountains and allows scientists to immerse themselves in the elements and their dreams. This chapter examines the excitement generated by the early balloon flights and the implications for the mapping of the terrain, both in topographical applications and for ethnographic studies.

Chapter Six explores how lithography was born out of the desire for rapid, large-scale and cheap reproduction of documents and pictures. Napoleon was interested in the process because he saw it as a way to ensure the speedy diffusion, during military campaigns, of facsimiles of orders and communiqués. Lithography was a technical and commercial proposition which was in turn linked to graphic reproduction that in the space of fifty years, from the early years of the nineteenth century, became the most powerful way to illustrate historical and social events, often appearing in newspaper form. Caricature plays a large part in these representations so the work of Daumier, in particular, is looked at in some detail. Baudelaire, Balzac, Guys, Courbet, Maistre and Manet, amongst others, also get a look in.

The scene of a diorama – the subject of Chapter Seven - is carefully arranged to suit the gaze of the spectator. With little effort on the part of a viewer, representations of ancient buildings, landscapes, city scapes and battles are presented in a well ordered space. Such scenes are designed to attract, in this case the European observer, while the risk taken encountering the exotic or dangerous, the 'other' is minimized. Daguerre for example contrived an eruption of Vesuvius with real rocks showering on the stage. Two other displays by Daguerre were the talk of the town in London and Paris in 1825 and 1826, 'Ruins of Holyrood' and 'Cathedral of Chartres.' These dioramas have a privileged place in the process of a rediscovery of the past, constituting self and other, representing individuals or sections of society or national characteristics and classifying in an enclosed space in a linear way. Daguerre is an enthusiastic conversationalist and showman.

Chapter Eight examines the early years of photography. Photography arrived at a specific historical juncture in the modern world. It promised to fulfil the desire to accurately represent nature and the social world, and, more than this, it is inseparable from the epistemological shift at the turn of the eighteenth century that brought with it the reorganisation of knowledge across a variety of fields, a change in social governing practices and changes in the desiring capacities of human beings. Photography itself arose out of a widespread desire to fix permanently the fleeting images seen through a camera obscura. Daguerre features in this chapter too, along with people like Arago, Bayard, Chevalier, Talbot and Niépce. Chapter Eight is rich in debate on this, at the time, contentious technology.

CHAPTER ONE

Perspective

I watch Alberti writing his 1436 dedication to his friend Brunelleschi at the front of the Tuscan translation of *Della Pittura*. He has been careful in his choice of words, ruing what he sees as lost from the ancient world, but marvelling at the endeavours of his friends and the possibility of them equalling and surpassing the enterprises of these 'venerated painters, sculptors, architects, musicians, geometers, rhetoricians, augers and suchlike distinguished and remarkable intellects'.[1] The quill scratches neatly forward on the parchment 'I then realized that the ability to achieve the highest distinction in any meritorious activity lies in our own industry and diligence no less than in the favours of nature and of the times.' He is optimistic and hardworking – convinced that it is his duty to cultivate *virtu* - and blessed by Nature, as a bulwark against the vagaries of *fortuna* and at this moment gripped by his first

---

[1] Alberti, L.B., *On Painting*, Penguin, England, 1972 (1436), p.34.

look at his friends just completed dome of Santa Maria del Fiore in Florence. He must mention it, 'What man, however hard of heart or jealous, would not praise Filippo the architect when he sees here such an enormous construction towering above the skies, vast enough to cover the entire Tuscan population with its shadow, and done without the aid of beams or elaborate wooden supports.'[2] He hasn't finished but seems to contradict his opening paragraph about Nature growing weary and no longer producing giant intellects 'Surely a feat of engineering, if I am not mistaken, that people did not believe possible these days and was probably unknown and unimaginable among the ancients.'[3] I interrupt his musing to challenge him with what I see as his romanticised vision of the ancient world that is somehow disconnected from the work he and his friends are doing in the present. He responds 'I applaud the ethical codes of the Roman authors, in particular the works of Cicero and the architect Vitruvius, though I radically reorder the

---

[2] Ibid., p.35.

[3] Gombrich makes the astute observation that Alberti actually paraphrases Pliny in his moving eulogy to Brunelleschi. See Gombrich, E.H., *Journal of the Warburg and Courtauld Institutes*, Vol.20 (1960), No 1-2 p.173.

presentation and underlying logic of his treatise.[4] What I am doing is bringing this important work to my contemporary audience'. 'And further', he says in response to my sceptical look, 'it is just as important to be regarded as a great imitator of the ancients as it is to be up to date and an innovator, and this little book will prove I am both.' 'Why Vitruvius?' I ask. 'Because it is the only text on the visual arts that has survived from antiquity,' he replies, 'and further because of the important things he has to say about geometric proportion in the human body and architecture'. I don't doubt his assertion about *Della Pittura* and need look no further into why Vitruvius.

The detailed work on architecture by the Roman Vitruvius had been discovered in 1410. Alberti had marked the first chapter of the third book where Vitruvius wrote,

> Without symmetry and proportion no temple can have a regular plan; that is, it must have an exact proportion worked out after the fashion of the

---

[4] In fact he did a lot more than this. His constructions are more Ciceronian than Vitruvian and he replaces Greek technical terms with others of Latin derivation. He is at pains to replace Greek influence with Roman, as his later writings on Architecture show. See 'Alberti and Filareth', John Onians, *Journal of the Warburg and Courtauld Institutes*, Vol.34 (1971), pp.96-114.

limbs of a finely shaped human body. For nature has so planned the human body that the face from the chin to the top of the forehead and the roots of the hair is a tenth part; also the palm of the hand from the wrist to the top of the middle finger is as much; the head from the chin to the crown, an eight part...In like manner the various parts of the temples ought to have dimensions answering suitably to the general sum of their whole magnitude.[5]

Vitruvius compiled his illustrated treatise *De architectura* in ten books partly from his own experience and partly from earlier Greek architects. His main interest was in preserving Hellenistic Greek practice – with its characterised loss of restraint and repose from earlier Greek art and a striving for theatrical effect - and his theory of human proportion, in which the human figure is shown to fit into a square circumscribed by a circle, has become known as 'Vitruvian man'.

Alberti is attentive to the day-to-day ways of his world, commerce, industry, invention, art, music and politics

---

[5] Cronin, V., *The Florentine Renaissance*, The Folio Society London, 2001 (Collins 1967), p.189. For the full description see Vitruvius, *The Ten Books on Architecture*, Trs. Morgan, M.H., Dover Publications, New York, 1960 (1914), pp.72-75.

(he travels around under a papal preferment and remuneration) and believes there is an underlying order, prescribed by God (Christian), on which the individual can base their social behaviour. I ask him to explain how he reconciles commerce and *virtu*, 'There are...activities in which the powers of body and mind function together to bring profit. Such are the occupations of painters, sculptors, musicians and others like them. All these ways of making a living, since they depend mainly on our personal powers, are what you call arts, and do not go down in shipwrecks but swim away with our naked selves. They keep us company all our lives and feed and maintain our name and fame.'[6] 'So,' I summarise, 'our stripped down selves are both body and mind, orchestrated by God through Nature, with incumbent creative powers that will be brought out by our hard work and saved by our swimming abilities'. He laughs at my joke, appreciating it as only someone who wrote a ringing eulogy of his faithful dog in order to mock the formulas and pomposities coming to dominate humanist writing could do, and he elaborates for my benefit 'My dog was' he says 'born of most noble ancestors...and amongst his ancient ancestors were an infinite number of famous

---

[6] *I Libre della Famiglia*, trs. Watkins, R.N. *The Family in Renaissance Florence*, Columbia (South Carolina), 1969, p.145.

persons,' including 'ancient and learned persons of Egypt'[7] Ah, my dog is better than yours.

'Alright', I tell him, 'I have some background but my questions for today have to do with perspective and the place of man in the scheme of things.' 'I need to finish my dedication to Brunelleschi first' he replies, 'then you will have an outline to formulate your questions from.' More parchment scratching as I note how he carefully forms the words, especially the capitals, using strict geometric formula, he also rules lines horizontally and vertically so his text doesn't wander on the page or out of the margins. Despite the scant punctuation (it only appeared with the printed texts in Italy in 1465), the result is pleasing to the eye,

> You will see that there are three books. The first, which is entirely mathematical, shows how this noble and beautiful art arises from roots within Nature herself. The second puts the art into the hands of the artist, distinguishes its parts and explains them all. The third instructs the artist

[7]'Canis' in *Apologhi ed elogi*, ed. Contarino, R., Genoa, 1984, p.142 and 162.

how he may and should attain complete mastery and understanding of the art of painting.[8]

He then writes a more personal appeal to Brunelleschi 'Please, therefore, read my work carefully, and if anything seems to you to need amendment, correct it.[9] He puts his goose quill down, carefully blots the page and looks up; 'I take from the mathematicians what is relevant to the subject. But in this matter I speak as a painter, not a mathematician.' He clarifies his position further 'Mathematicians measure the shapes and forms of things in the mind alone and divorced entirely from matter. We, on the other hand, who wish to talk of things that are visible, will express ourselves in cruder terms.'[10] 'Are you saying that mind and matter are not connected and that mathematics is a higher form of thought?' 'No' he explains, 'When I say cruder I am referring to a Ciceronian expression *Minerva* which means the coarser wisdom of our senses.' 'That seems to me a curious thing to say and is out of accord with what you have said to this point' I respond, 'unless you are referring more to the Italian goddess of crafts and trade guilds than the Greek

---

[8]Alberti, L.B., *On Painting*, p.35.

[9] Ibid.

[10] Ibid., p.37.

Athena'. 'What I am doing is keeping the emphasis on sensate knowledge and I do this by using a series of material metaphors and similes to describe the geometrical properties of bodies'[11] 'and' he continues, 'Don't forget my purpose is to write a book that can be understood by people who are not mathematicians, young up and coming artists, people who are dealing with visual descriptions of the material world. This is also the reason I am translating it to Tuscan from Latin.' 'What about mathematics being a higher form of thought?' I remind him. 'You need to read more carefully' he smilingly quips 'the virtues of painting...are that its masters see their works admired and feel themselves to be almost like the creator. Is it not true that painting is mistress of all the arts or their principal ornament?'[12] Alberti goes on to explain how painting possesses a truly divine power that makes the absent present and represents the dead to the living. How it has been a great gift to men in the way it represents the gods they worship, binding us to the gods and filling our minds with sound religious beliefs.[13] He refers to people like Plutarch and

---

[11] Kemp, introduction, Alberti, L.B., *On Painting*, p.17.

[12] Alberti, L.B., *On Painting*, p.61.

[13] Ibid., p.60.

the stories he told about the effect Alexander's portrait had on one of his commanders to drive home his point. I understand then mathematics does not take a higher form than God but I am not sure I articulated my question about mathematics being divorced from matter well enough. Or perhaps he didn't answer directly. I decide not to question him about how he so easily slips between description of God and the gods and just take it that he is so influenced by the text of Vitruvius he forgets himself. As Alberti talks a question I have been turning over in my mind clarifies. It has to do with what he says about treating human subjects in the important *historia* paintings.

'If I have understood you correctly' I begin, 'your sole purpose in *Della Pittura* is to establish the rudiments of optical geometry external to the eye, and their consequences for the painter.'[14] 'Yes' he nods. 'That would explain why' I continue 'that despite being well read in and animated by medieval optical science, including the work of people like the Islamic philosopher Alhazen, the Franciscans Roger Bacon and John Pecham, and Witelo from Poland,[15] you contain yourself

---

[14] Kemp, introduction, Alberti, L.B., *On Painting*, p.12.

[15] Ibid., p.11.

to a visual description of the material world and obliquely refer to 'those philosophers who are experts.' 'That is right' he replies, 'and you must remember that the science provided by these people offers a perfect illustration of the way in which God's rational design insinuates itself throughout nature.'[16] 'We in Florence, since Dante, have followed Plato in order to make a more serious and fruitful use of number. The mechanical clock, newly arrived, has hands that translate organic time into mathematical units on a dial. The second innovation to have just arrived and based again on strict mathematical principles, is perspective. This introduction allows us to represent objects in space.'[17] Alberti smiles 'being practical we left aside, as you can see, Plato's doctrine that physical phenomena are not in themselves worth studying.'[18] I pick up on his use of the word introduction, 'this is all part of the classical revival which began in sculpture then began to influence architecture and recently painting.' 'That's right' he replies. 'Apparently there are a few classical pictures that survived but none that we in Florence know about or

---

[16] Ibid.

[17] Cronin, V., *The Florentine Renaissance*, p.127.

[18] Ibid.

have seen. The writings of Vitruvius have become more crucial because of this lack.' 'Apart from the Vitruvian man' I continue 'Vitruvius also writes about perspective doesn't he?' Alberti hesitates before answering 'as we can judge from the works of former ages, this matter probably remained completely unknown to our ancestors because of its obscurity and difficulty.' 'And' he continues 'you will hardly find any *historia* of the ancients properly composed either in painting or modelling or sculpture.'[19] 'That may be' I counter 'and I accept you don't actually have paintings to look at, but Vitruvius wrote about perspective when describing set construction for theatre. He was talking about Agatharcus, and I read 'at the time when Aeschylus taught at Athens the rules of tragic poetry, was the first who contrived scenery, upon which subject he left a treatise. This led Democritus and Anaxagoras, who wrote thereon, to explain how the points of sight and distance ought to guide the lines, as in nature, to a centre, so that by means of pictorial deception the real appearances of buildings appear on stage: painted on a flat vertical surface, they seem nevertheless to advance and recede.'[20] 'That may be so,

---

[19] Alberti, L.B., *On Painting*, p.58.

[20] Cronin, V., *The Florentine Renaissance*, p.200.

but what we have discovered now, and have demonstrated to our friends who have described the results as 'miracles of painting', has to do with my centric point, and the construction of triangles, the pyramid and the intersection.' 'But' I respond 'what you write about the drawing of geometrical figures on a foreshortened pavement does not meet the need to show how the pyramid results in the pictorial construction.'[21] 'I don't have time now and I used to demonstrate these things at greater length to my friends with some geometrical explanation. Here I am giving the rudiments to unlearned painters.'[22]

The reason Alberti won't answer my question and resorts to an unsatisfactory explanation is because the perspective he is writing about is being practiced by his friend Brunelleschi and the painter Masaccio (rendering architecture but not figures, at least on his *Trinity*). It is not clear Alberti is claiming he invented it; just that he is the first to write about it.[23] This may explain why he

---

[21] Alberti, L.B., *On Painting*, p.98, notes.

[22] Ibid.p.59.

[23] Some 40 treatises appeared before 1600. Together these treatises contain hundreds of individual methods only a tiny percentage of which repeat Alberti's (and usually in a mistaken manner). In short the field is huge. See 'Response to Tomas Garcia Salgardo', James Elkins, in *Leonardo*, Vol.29, No.1, 1996, pp.82-83.

doesn't offer a workable description. Of course by the time Vasari wrote in his 1568 second addition of *Lives* that Alberti 'discovered a way of tracing natural perspectives and effecting diminution of figures, as well as a method of reproducing small objects on a larger scale: these were very ingenious and fascinating discoveries, of great value for the purposes of art,' it was authoritative text.[24] Alberti doesn't mention Ibn-Al-Haithan (Alhazen) but his writings on perspective were well known and the copy of his work on optics in the Vatican has been annotated by Lorenzo Ghiberti which, along with his knowledge of Brunelleschi's work, may account for his use of perspective when sculpting the reliefs for the doors of the Baptistry, firstly the Northern doors (1401-1424) then the Eastern doors (1425-1452). Directly acknowledging Ibn-Al-Haithan would go against one of the main aims of Alberti, to place Florence in a direct line with Roman heritage, thus removing or diminishing the Greek roots, along with any Eastern influence. He strongly argues against Egyptian influence.[25] Of course anyone who has visited the Piazza and Basilica di San Marco in Venice can't ignore the

---

[24] Vasari, G., *Lives of the Artists*, The Folio Society, London, 1993(1965 – first published 1550 and enlarged 1568, p.131.

[25] Alberti, L.B., *On Painting*, p.61.

obvious influence of the East. I will leave Alberti to his writing for the moment but will return soon to ask him some questions about Protagoras and his descriptions of composition in *historia*.

Brunelleschi arrived at an understanding of perspective by reading the text of Vitruvius closely and applying it in a practical way. The exterior of the Baptistery, the building dedicated to St John the Baptist, patron saint of Florence, probably the oldest structure in Florence and of Roman origins, with its squares of dark green marble inlaid on white receding down two sides, provides a singularly dramatic illustration of Vitruvius' text.[26] I catch up with Brunelleschi at around midday as he is making an accurate drawing of this phenomenon as seen from the centre door of the cathedral Santa Maria Del Fiore. I stand next to him on the marble steps and watch as he draws on a thick wooden panel nine inches square.[27] Animated by his sense of theatre he extends the drawing to give the illusion of three dimensions. In order to achieve this he has invented, he says, a new method of looking at such drawings. The description he gives

---

[26] Cronin, V., *The Florentine Renaissance*, p.200.

[27] Ibid.

sounds a bit like a camera obscura, complete with mirror and I ask him to show me. 'I have pierced a funnel-shaped hole in my panel with the wider opening at the back. To view it I put my eye to the back of the hole like this and hold a mirror nine inches square at arm's length, facing the picture. What I can see in the mirror is a stereoscopic view of the Baptistery, with the squares of dark green marble receding regularly, and the shops of Piazza San Giovanni behind.' He hands the painted panel and mirror to me to try. As I marvel he explains the importance of what has just happened 'This has changed our conception of space. Rather than space being treated as something in front of the subject like the Byzantines had done, we Florentines see it as something behind.'[28] The panel and mirror is like looking through a window at the scene depicted, the gaze is directed right into the picture by receding lines and shapes. 'As you can see' continues Brunelleschi 'my painting is exactly as Agatharcus described. It is the same as a stage back drop that creates the illusion of three dimensions.'[29] 'What is the next step?' I ask. 'I superimpose the perspective lines on my panel. I also draw directly on to the mirror for further

---

[28] Ibid., p.201.

[29] Ibid.

effect. I have detailed ground and building plans to complete the picture.' 'So you are not applying the geometrical rules of perspective in order to do your drawing?' 'No, I haven't learned the art of mathematics yet, but my friends are taking me to see Paolo soon.'[30] Now I understand why Alberti had trouble describing the process to me but at the same time cannot imagine that Brunelleschi would not have demonstrated his invention to his friend. Perhaps when Alberti talks about 'miracles of painting' or his 'drawing boxes' he is referring to a camera obscura. He certainly gives full description to his 'intersection' or veil as an aid to drawing.[31] But I will come back to this question further along. Above all Brunelleschi is a practical man, a craftsman, a man who makes, says Vasari, 'some very splendid and beautiful clocks,'[32] and crucially he builds three dimensional models that occupy space and offer all-round vision. At the same time his good friend Donatello frees sculpture from its Gothic influence by carefully studying nature and allowing humanity – a complete physical and spiritual being - to emerge and I am back in the moments

---

[30] Vasari, G., *Lives of the Artists*, p.77.

[31] Alberti, L.B., *On Painting*, p.98, note 15 and pp.65-66.

[32] Vasari, G., *Lives of the Artists*, p.133.

I need to be in order to discover the underpinnings to seeing the social.

When I catch up with Alberti I tell him I have seen Brunelleschi's mirror and panel. He is quiet for a minute then admits 'I have been working back to front as a way of proving my thesis on perspective, without actually applying the theory to practice. Brunelleschi, by contrast not only drew the Baptistery but has also drawn the Piazza San Giovanni showing all the squares in black-and-white marble receding beautifully, and he also drew in the same way the house of the Misericordia, with the shops of the wafer-makers and the arch of the Pecori, and the pillar of St Zenobius on the other side.[33] You will have noticed there are no illustrations in my little book.' His enthusiasm for the drawings of Brunelleschi is infectious and I am glad he has told me about there being no illustrations. My earlier readings were taken up with drawing long bows on vision with a disembodied eye sending out centric, median and extrinsic rays to strike points on a plane.

'Alberti', I proffer, 'You often quote Protagoras in your writings and use him in various ways,' 'Ah yes' he replies,

---

[33] Ibid., p.76.

'In *Della Pittura* I say as man is the best known of all things to man, perhaps Protagoras, in saying that man is the scale and measure of all things, meant that accidents in all things are duly compared to and known by the accidents in man.'[34] 'So you are in line with Protagoras and the Medieval opticians we mentioned above, as you regard the figure of man to be the most standard of reference' he nods assent.

Unfortunately the major works of Protagoras like his *On Truth* and *On the Gods* have not survived and we only know his writings through Plato who quoted the maxim 'a person is the measure of all things' and has him appear in *Theaetetus* in dialogue with Socrates, Hippias and Prodicus. It is worth keeping sceptical eyes open when one philosopher quotes another, especially when introduced into the dialogue of a supreme puppet master like Plato. Of Protagoras' ipsissima verba (actual words, as opposed to paraphrases), the most famous is the homo-mensura (man-measure) statement (DK80b1): 'Of all things the measure is man, of the things that are, that [or 'how'] they are, and of things that are not, that [or 'how'] they are not.' This precise meaning of this

---

[34] Alberti, L.B., *On Painting*, p.53.

statement, like that of any short extract taken out of context, is far from obvious, although the long discussion of it in Plato's *Theaetetus* gives us some sense of how ancient Greek audiences interpreted it.[35]

Protagoras was sceptical about any claim to absolute and universal truth and taught a doctrine of relativity in all knowledge. It could be argued that what he was saying was *each* man is the measure of all things, so there is no objective truth'[36] It is not clear from what Alberti says if this means man is used as the scale, and there are plenty of illustrations and description of the figure man given to support this view, or that man is an unreliable or capricious individual who requires forgiveness, but taking the first salient part yes man is the measure of all things – pictorially how Alberti presents it is important to my story - he seems to do both. 'You seem to qualify your point about Protagoras and reduce it to pictorial composition', 'Yes, however small you paint the objects in a painting, they will seem large or small according to the size of any man in the picture.'[37] 'By any man you mean

[35] http://www.iep.utm.edu/p/protagor.htm#SH3b

[36] Russell, B., *A History of Western Philosophy*, George Allen and Unwin Ltd., London, 1946, p.97.

[37] Alberti, L.B., *On Painting*, p.53.

your modular man?' 'Of course.' This brings me back to 'man is the measure' and what Alberti writes in Book 2 when he begins describing composition in *historia*. 'You are at great pains to introduce conformity and harmony and order and you do this by advocating an ideal, in fact you go further and complain what real people in the world look like should not appear as they are because of sensibility and design. In a similar way to Brunelleschi criticising Donatello saying he had put the body of a peasant on the cross not the perfect form of Jesus'[38] 'You seem to keep forgetting what it is I am trying to do in my little book' he protests.

My problem is that although Alberti keeps insisting his little book is a textbook for young artists it clearly does not give practical instructions on how to make paintings, whether construction, form, line, colour,[39] perspective, materials, sourcing of materials, classes...he is doing

[38] Vasari, G., *Lives of the Artists*, p.98. Donatello employed optical corrections to his sculpture. These were used to compensate for the artists and subsequent spectators viewpoint - either above or below the sculpture. See 'Optical Corrections in the Sculpture of Donatello', Robert Munman, in Transactions of the American Philosophical Association, Vol.75. Part 2, 1985. Similarly Masaccio's *Trinity* is best viewed from a set distance (about fourteen feet) and the spectator should kneel. Of course Masaccio applied the rules of perspective to the architecture, not the figures, which remain on a flat plain.

[39] 'Alberti's Colour Theory: A Medieval Bottle without Renaissance Wine,' Samuel Y. Edgerton, Jr. Journal of the Warburg and Courtauld Institutes Vol. 32 (1969), pp. 109-134, for the colour aspects to *Della Pittura*.

something different - not just in a hurry to get the palm for being the first to write about perspective – and the key is his attitude to *historia*. A closer reading of his text just seems to take me further away from understanding what he is saying. He shifts the ground from composition and comparative figure modular man to thought and knowledge.

'Alberti', I try again 'You use Quintilian's description of a [now lost] marble relief depicting Meleager from ancient Rome as your prime example of historia and use his words and rhetorical style in your text.' *Institutio* was discovered by Poggio Bracciolini in Switzerland in 1416 and was described by Martialis as 'the supreme guide of wayward youth'[40] Poggio copied the entire work in 32 days.[41] 'Yes' he replies 'Composition is the procedure in painting whereby the parts are composed together in the picture. The great work of the painter is not a colossus but a 'historia', for there is far more merit in a 'historia' than in a colossus'[42] He notices I am still puzzled 'Parts of the 'historia' are the bodies, part of the body is the

[40] *The Oxford Companion to Classical Literature*, Howatson, M.C. (Ed), Oxford University Press, Oxford, 1989, p.478.

[41] Cronin, V., *The Florentine Renaissance*, p.35.

[42] Alberti, L.B., *On Painting*, p.71.

member, and part of the member is the surface.'[43] He continues in this vein for a few minutes but he has lost me. He has a circular explanation going that has no definition embedded. I am searching for another question when I catch his sentence 'From the composition of surfaces arises that elegant harmony and grace in bodies, which they call beauty.'[44] 'So you advocate going beyond artistic talent in representation towards harmony, beauty, expression, symmetry and significance.' 'You have it. The central purpose of all my work is the cultivation of *virtu* and *Della Pittura* is designed to pass this knowledge on to the young.' 'I believe, using Pliny's *symmetria* as my guide, that is in a standing person [you] will note the whole appearance and posture, and there will be no part whose function and symmetry, as the Greeks call it, [you] will not know.'[45] 'But' he continues 'considering all these parts, [you] should be attentive not only to the likeness of things but also and especially to beauty, for in a painting it is as pleasing as it is necessary.'[46] Quintilian gives a good example of this

---

[43] Ibid.

[44] Ibid.

[45] Ibid., p.90.

[46] Ibid.

when discussing the painter Demetrius. I have been looking for tensions in Alberti's discussion that he doesn't see or agree with. This could be because the tensions are mine or because what he is saying he is doing is not what his descriptions show me. Some of it comes about because I am looking for indications of seeing the social in his thinking and he isn't explicitly examining the 'social'. In fact it is unlikely he would know what I am talking about when I discuss the social.

Tacitus's (Cornelius) *Historiae,* sections of which survived, was part of a body of work that encouraged Florentines to bind themselves in spiritual alliance to republican Rome. Leonardo Bruni quotes a part of Tacitus's work where he is discussing historiography. It was Bruni who used the phrase *studia humanitatis,* meaning the study of human endeavours versus those of theology and metaphysics, which is where the term humanists comes from. His *History of the Florentine People* in twelve books (official publication 1442) made his work the best seller in the fifteenth century. As a humanist Bruni was essential in translating many works of Plato and Aristotle. His use of Aelius Aristides' *Panathenicus (Panegyric to Athens)* to buttress his republican theses in the *Panegyric to the City of Florence* (c. 1403) was instrumental in bringing the Greek historian to the attention of Renaissance political philosophers.[47] Although Alberti doesn't refer to Bruni directly it is likely he knew his work.[48]

Alberti's *historia* figure is at once something more and something less than a 'real' man, a 'social' being, at once recognised, but rather a revered or mythical being with individual difference and distortion and disfigurement removed, a homogenised or amalgamated figure, almost like a homunculus. Alberti mentions in his last essay that he would like painters to paint his portrait in their 'historiae' as recognition of his labours. The title *Historiae* was used by Tacitus to describe the period AD 69-96. For later academics historia – a word that resists direct translation - means roughly 'history painting', a human narrative drawn from a significant secular or Christian story or in Alberti's case allegorical representation.[49] It is

---

[47] See Hans Baron's *The Crisis of the Early Italian Renaissance,* Princeton University Press, Princeton, 1955, for details.

[48] For the closest comparison see 'The Languages of Literature in Renaissance Italy' by Peter Hainsworth; Valerio Lucchesi; Christina Roaf; David Robey; J.R. Woodhouse. Reviewed in ITALICA, Vol. 67, No. 3 (Autumn, 1990), pp. 400-403. For the importance of Bruni see *The Crisis of the Early Italian Renaissance. Civic Humanism and Republican Liberty in an Age of Classicism and Tyranny* in two volumes by Hans Baron, Princeton University Press, Princeton, 1955, and his later *In Search of Florentine Humanism: Essays on the Transition from Medieval to Modern Thought,* Princeton University Press, Princeton, 1988.

[49] Alberti, L.B., *On Painting*, p.99, notes.

also interpreted as giving voice to humanistic design with all its philosophical and historical range. Alberti defines a pictorial order that is also an order of thought and knowledge: the Representation, a modular man whose influence would last several centuries. Finally *historia* has been described as dramatic action and composition connected, thus pulling the emotional chords of the hypothetical spectator - the spectator who recognises aspects of themselves but not as a likeness. Alberti himself has more to say about what this figure is not, and clearly man is not a recognizable figure, an individual or unique being of the world. Man is rather a representation of beauty and symmetry. To me this is at odds with the self-portrait Ghiberti placed in a central position on his 'Paradise' doors of the Baptistry, or Donatello's sculpture of the bald headed prophet Abacu or his emaciated Magdalene. However Alberti's representation has implications for later seeing the social because *Della Pittura* was, as he designed, so influential. It could be assumed Alberti would prefer painters make a likeness of his image, as he self-portrayed, rather than a *historia* figure.

'Well Alberti, that will do, I know you are a busy man and that your little book, as you call it, is only one small contribution to the vast writing across many fields that

occupies your time. Do you have a final few words to say?'
'All who wish their works to be pleasing and acceptable to posterity, should first think well about what they are going to do, and then carry it out with great diligence. Indeed diligence is no less welcome than native ability in many things. But one should avoid the excessive scruple of those who, out of desire for their work to be completely free from all defect and highly polished, have it worn out by age before it is finished...wanting to achieve in every particular more than is possible or suitable is characteristic of a stubborn, not of a diligent man.'[50] The preparation of a painting is lengthy and thought through, along with the topic and the composition. The execution begins at the right moment and is completed at the right moment and crucially it is not laboured. The finished work should represent harmony, beauty, expression, symmetry and significance. Above all the process of painting, as in life, is to cultivate *virtu* in order to use it as a defence against the vagaries of *Fortuna*.

---

[50] Ibid., pp.94-95.

CHAPTER TWO

Lenses

If I was writing a chronological history of the use of lenses I would start with luminaries like Euclid, Seneca, Amarti, Spina, Redi, Ibn-Al-Haithan, Roger Bacon, da Vinci, Maurelico, Descartes and Kircher and so on, but this is not my purpose, and it has already been done, both fairly and with the incumbent historiographical trends and nationalistic prejudices. What I do instead is pick out texts that help describe how the microscopic, and to a lesser degree the macroscopic lenses, make possible a vision of the social. The best way to describe the experience is to liken it to entering a vast hall of luminary and amateur and weekend scientists and others who are fascinated by the natural world and the possibilities opening up to seeing the minute and infinite (and without the aid of a lens an invisible) world they inhabit. What I will do is wander around listening to snatches of conversations, joining some that I find useful and look at the various displays and publications on show. The only forward time limit on the gathering is 1876 and Robert

Koch, chosen because it is a turning point that elevates microscopy to the centre of the scientific and social stage; but mostly I stick with the early discoveries. The analogy of the crowded hall is carefully chosen. It is to make clear that this history of lenses and the advances made is not a history of brilliant individuals working in isolation and in competition with other individuals but it is a shared learning. The hall is a vast clearing house of ideas and improvements, with little concealment, apart from the occasional like Leeuwenhoek's secrecy about his unique viewing style that he guarded until his death, then afterwards, with speculation the guide to how he achieved his results...but I can ask him in a few minutes. He was most keen to be here. This gathering is not however a scientific community that shares a common goal or belief as we would understand it like a physics faculty in a university dedicated to solid-state studies. In the few minutes they are gathered together the participants are concerned about lenses, their uses and meanings, the creative process, the technicalities but that is all. For them microscopes and telescopes are only a small part of other ongoing enterprises, some of which are scientific as we would understand it.

I shift my position slightly so that the glass from the display case doesn't reflect my image. I almost have the room to myself which is a boon since I can observe and write and draw in my notebook unselfconsciously. The Science Museum in London has some examples of early Italian compound microscopes. At approximately 3-5 inches in height and about the thickness of a thumb they are quite small. There are no examples of the early Dutch microscopes. This is curious because Leeuwenhoek corresponded regularly with the publishers of *Philosophical Transactions* and many of his observations are recorded in its pages. Two years after his death 247 microscopes and 172 lenses mounted between plates were sold at auction. Apparently none have survived, including the 26 little instruments that were sent to the Royal Society where they were cared for until 1820. The last record of them suggests that a celebrated surgeon Everard Home took them home.[51] By copying drawings an enterprising person has made a replica of one of Leeuwenhoek's microscopes. It too is about the size of a thumb and it has a flat metallic piece with a winding screw mechanism. If it didn't have a caption I wouldn't pick it as a microscope - it looks more like a tiny paddle. It has a single lens – remembering Leeuwenhoek made a new lens for each specimen – and the object to be viewed

---

[51] Ford, B.J., *The Revealing Lens: Mankind and the Microscope*, Harrap, London, 1973, pp.62-63.

is mounted on a pin at just the right distance.[52] To view small objects such as the eating apparatus of a louse would take great skill and patience, two very steady hands and crucially good light. Leeuwenhoek had some tricks up his secretive sleeve though and I will come to them when I examine more closely his work.

The Musschenbroek workshop, De Oosterse Lamp (The Oriental Lamp), has a display here. The Musschenbroek's migrated to Leiden from south Netherlands around 1610 and began a brass foundry. One of their specialities was a small domestic oil lamp and they named the shop after this product. Fortuitously the shop is near the University of Leiden and the Musschenbroek's have begun to manufacture scientific and medical instruments, including microscopes. On a table of the display are a range of the unique trade catalogues they produce. A browse shows they are producing hundreds of different scientific instruments and usefully the prices are included in the catalogue. I am not the only person looking at this display and rub shoulders with professional scientists, university people and amateurs from all over the world. Apparently a little earlier in the

[52] Hacking, I., *Representing and Intervening*, Cambridge University Press, Cambridge, 1983, p.192.

day the Russian Czar Peter the Great and the Dutch Prince of Orange had been here.[53] The Czar purchased two microscopes and a large Senguerd-type air-pump on its oak pediment, with twenty accessories.[54]

I spot Copernicus in earnest discussion with Calvin. I had forgotten Copernicus was a high ranking Polish clergyman (he has a degree in canon law and medicine). I always picture him as an astronomer gazing alternatively at the night sky and at the writings of Ptolemy, so his appearance jolts me. In one hand he has his quadrant, in the other *De revolutionibus orbium coelestium*, published in Nuremberg 1543. In his book he wrote that the earth is not fixed at the centre of the universe but rather it is a planet that moves around the sun. He does his research using astronomical instruments like the astrolabe, the quadrant he holds, the triquetrum (a parallactic ruler), and an armillary sphere (a model of the celestial sphere). He does not have a telescope because they are not yet invented. He began by comparing the tables of Ptolemy and Alfonso with his

---

[53] de Clercq, P., *At the Sign of the Oriental Lamp: The Musschenbroek Workshop in Leiden. 1660-1750* This study may be considered as a companion volume to the catalogue The Leiden Cabinet of Physics, published by the Museum Boerhaave, the Dutch National Museum of the History of Science and Medicine.

[54] Ibid., p.165.

own careful observations – and those of his friends - for over fifty years.[55] The fact that Copernicus holds an important post in the Roman Catholic Church presents a paradox for a history of science that relies on a strict division between science and religion to tell its story, the construction of a 'straw man'. I overhear Calvin saying 'No I am not anti-Copernican. I know there have been many scholars who have credited the sentence *who will venture to place the authority of Copernicus above that of the Holy Spirit*, to me but I didn't say that.'[56] 'I did however' he continued 'speak one harsh passage against you in a sermon in 1556,

> We shall see them, so frantic not only towards religion but also for showing everyone they have a monstrous nature, who will tell the sun does not move, it is the earth which is active and turn. When we see such spirits, we must tell ourselves the devil

[55] *The Life of Copernicus (1473-1543)*, Pierre Gassendi with notes by Olivier Thill, Xulon Press, Fairfax, USA, 2002.

[56] See Rosen, E. 'Calvin's Attitude towards Copernicus' in *Journal of the History of Ideas*, New York, 1960, pp.431-441.

possess them and God proposes them to us as mirrors, in order to let us stay in His awe.[57]

Drawing breath he follows 'I actually think astronomy is very useful. In a sermon on the book of Job I said 'Therefore, clever men who expend their labour upon it are to be praised and those who have ability and leisure ought not to neglect work of that kind.'[58] 'But you are not a Copernican' proffered Copernicus. A definitive 'No' is Calvin's retort. I was contemplating the mirror in Calvin's sermon. We generally look at ourselves in a mirror, not at other people. I was also thinking perhaps it is theological differences between the two men that might be rewarding to examine, but not here.

Gassendi nudges me on the arm and tells me that Copernicus did not make the quadrant he is holding. 'Despite what others may have told you Copernicus bought his instrument. He comes from a very rich family and never learned manual labour. He also gives his books to a craftsman to bind.' 'What about the other astronomers?' I ask, 'Except for Galileo and Huygens,

---

[57] *Calvini opera*, Vol.49, page 677, Sermon 8 on the first letter to the Corinthians, preached in 1556, quoted in *The Life of Copernicus (1473-1543)*, Pierre Gassendi with notes by Olivier Thill, p.220.

[58] Ibid., p.219.

every astronomer buys rather than makes their instruments' he replies.[59] 'It might surprise you to hear Spinoza the philosopher from Amsterdam grinds lenses for a living, but he is not very much interested in astronomy. Huygens makes his own telescopes, but he does not sell them.'[60] I confirmed with him I heard Galileo had a worker named Marcantonio Mazzoleni to help him in lens manufacture. Brahe had the German goldsmith and instrument maker Hans Crol stay on his island Hven.[61] It was apparent most other scientists and amateurs bought their instruments cheaply from the abundant craftsmen available, like the Musschenbroek workshop. I am presented with a picture of a thriving industry, energy, a sense of community and friendship when preparing lenses, tooling up and viewing, even in these early days. The University of Cracow had two chairs of astronomy in Copernicus's time. Other important figures in the astronomy school were Martin Krol, Martin Bylica (who commissioned Hans Dorn to make celestial globes and other astronomical instruments and had an excellent library) and Adelburt Brudzewski, who taught

---

[59] Ibid., p.118.

[60] Ibid.

[61] Christianson, J.R., *On Tycho's Island*, Cambridge University Press, 2003, p.98.

Copernicus. Assuredly astronomy is not an individual activity. It is much more enjoyable to watch the sky with friends than alone.[62]

Museo del Castello in Milan houses one of the famous geometric compasses created by Galileo and built by Mazzoleni in 1606. Another is housed in a museum in Florence. Gemma Frisius also applied his mathematical expertise to geography, astronomy and map making. In Louvain he cooperated with the engraver and goldsmith Gaspard Van der Heyden (also known as Gaspar à Myrica) in the construction of maps, globes and astronomical instruments. His first publication was in 1529 when he produced a corrected version of Apianus's *Cosmographia*. The original had been published in 1524 and Gemma's new version, which stated on the title page that it was...*carefully corrected and with all errors set to right by Gemma Frisius*...was certainly a completely new printing by the Antwerp publisher Roeland Bollaert.[63]

In a quiet moment I notice there are three copies of Galileo's prohibited book *Dialogo* on display. I am not so

[62] *The Life of Copernicus (1473-1543)*, Pierre Gassendi with notes by Olivier Thill, p.181.

[63] http://www-history.mcs.st-andrews.ac.uk/Biographies/Gemma_Frisius.html

much interested in the text at the moment but rather struck by the different frontpiece etchings. In the first Italian edition of 1632 Aristotle, Ptolemy and Copernicus are engaged in animated discussion, standing in the open. Aristotle is most strident, chin jutting out and right index finger pointing, while Ptolemy listens with head inclined and Copernicus, palm upward seems to be responding. In his left hand - almost apologetically by his side - is his heliocentric model, bright sun in the centre. All three are bearded. In the Latin edition of 1635 Aristotle continues to hold court, his right hand straight pointing at the armillary sphere Ptolemy is holding. Ptolemy himself seems to be listening rather than speaking. Copernicus, who has had a shave, continues to hold his right hand palm upward while his model in his left hand is inclined upward. Copernicus is just looking at us, the spectators, leaving Aristotle and Ptolemy to their discussion. By the 1700 Latin edition the men are gathered in a Greek temple. Aristotle has aged and supports himself in a seated position. His upturned left hand seems to be pointing more towards Copernicus. Ptolemy, standing in the shade, is looking at Aristotle speaking while holding his model aloft, but he too is pointing at Copernicus's now eye level model with his left hand. Copernicus has two models in hand, left palm upward in a gesture of explanation. He has become younger again and continues to look out of the picture. But this time he is looking to the left of the frame and not directly at the spectator. As in the second picture his hat is gone along with his beard. I would like to ask the three illustrator's how much licence and how much direction they were given for the etchings but I can't find them. The rise in importance of Copernicus over the intervening 70 years seems evident.

Spinoza is not someone I would have expected to be here and no doubt he kept his head down when the Dutch Prince of Orange visited the exhibition, having sided with the De Witt's against the House of Orange. Spinoza has lived in Amsterdam and now The Hague all his life; his family escaping the Spanish Inquisition for The Netherlands. A man of simple wants he seems to earn a comfortable living from lens-grinding and with the support of small but regular donations from his friends. It is said that he makes excellent magnifying glasses, and some even suggest he is an optician. These aspects of his activities presents a curious conundrum given his devaluation of sense perception as a means of acquiring knowledge. His description is of a purely intellectual form of cognition and he idealizes geometry to use as a model for philosophy. For Spinoza sense perception has its origin in the action of an external body upon one or another of the sensory organs of one's own body. For example in his Ethics he says 'Different men may be differently affected by the same object, and the same man may be differently affected at different times by the same object.'[64] What Spinoza is trying to do is maintain a place for reverence and a life devoted to good works within the materialistic and deterministic physics of Descartes and his contemporaries. He constantly coughs then spits discretely into a handkerchief. I worry about that cough and wonder if it has something to do with breathing in the glass dust from the lenses he is grinding. At present he is working on a political thesis but I would really like to talk with him about rainbows.

Joris Hoefnagel and his sons Jakob and Johann are displaying their various publications including the fifty plates of insects engraved on copper and the 1592 paintings done on vellum long-titled *Archetypes and verses by Joris Hoefnagel, his father, are presented, engraved in copper under the guidance of his genius, and freely communicated in friendship to all lovers of the Muses by his son Jacob,* and the beautiful scarab beetle

---

[64] *Ethics* by Benedict de Spinoza (1677) Translated from the Latin by R.H.M. Elwes (1883) *MTSU Philosophy WebWorks* Hypertext Edition © 1997.

and other insects in pen and brown ink, coloured washes and gold paint on vellum by Joris is also here. Joris is a painter, poet, miniaturist and topographer. He travels widely and has produced nearly 100 views for Braun and Hogenberg's *Civitates Orbis Terrarum* of 1570-1618, and executed a plan of Cadiz for Ortelius, demonstrating that he has an eye for the macroscopic and the microscopic world. Jakob revises many of his father's plans for Braun and Hogenberg's Vienna publication of 1609. Johann is principally an engraver.

I spend a few minutes looking at Braun and Hogenberg's *Civitates Orbis Terrarum* because of its significance and because of the stories of discovery, chronicle and change they tell. Braun and Hogenberg worked for over twenty years to produce their 'towns of the world,' the first systematic views of cities and towns of the known world. There are six volumes to their enterprise, regarded as emblematic of Renaissance learning and designed to complement Ortelius' *Theatrum Orbis Terrarum* which is considered the first modern atlas. These two publications, as well as being works in progress, are responding to a burgeoning interest in the nature of the world and travel. Larger social movements like the emerging middle class with disposable money, combined with the availability of the printed book, assured widespread influence. Some evidence of this influence can be gleaned from the print runs of, for example Ortelius' atlas with the plates of Zelandicarum, Hollandiae Catth., Regni Hispaniae and Portugalliae, along with others, running into 8175 copies.

Three prints done by Joris for Braun and Hogenberg's *Civitates Orbis Terrarum* manage to catch my eye. I have to remind myself that the 'birds eye' views are done without the benefit of balloon flight, at best a high vantage point offered by a hill, but otherwise these scenes were drawn through careful cartographic and topographic study and detailed illustration. The first is a view of the small town of Alhama, southwest of Granada, drawn in 1563. Alhama is shown across a valley in mountainous countryside and obviously viewed from a natural high point. There are a number of local inhabitants gathered on a pathway in the foreground, chatting and walking, and a young girl is running. Someone else is riding a donkey. Everyone looks well dressed, well fed and content. The second print is a bird's eye view of the Dutch city Delft in 1588. The city is protected on all sides by a river or canal and high ramparts with towers and two guarded water gates. There are three main churches and possibly thirteen smaller ones. There are numerous windmills, and canals intersect the tree lined streets. There are a number of substantial buildings and some less so. There are also open areas or parks with plenty of carefully spaced trees. Around the periphery cows are grazing, a couple of farm buildings sit, and what looks like an important couple then a group of three equally important looking people are portrayed on either side in the foreground. If I had time I could count the number of houses and estimate the population of this ordered and safe city. The last print is a dramatic image of the city of Tunis along the Barbary Coast drawn in 1574. Again Joris has presented a bird's eye view, this time looking down from the north – a compass in the lake indicates - with the city, its waterways and fortifications clearly laid out below. Of historical significance, along with the social and cultural aspects, is that this hand coloured print shows the siege of Tunis by Turkish forces in 1574, the same year it was drawn. When issued it would have provided its viewers with a contemporary image of an important event that occurred south across the Mediterranean almost in newspaper form. The siege ended with the Spanish forced out of Tunis, which then became Turkish regency. The mechanics of the siege are clearly visible, ships in the harbour, troops on the ground in strategic positions and bivouacked in tents. Two caption boxes provide the commentary.

Joris's *Archetypa studiaque patrisä* of 1592 in four volumes is also here. While most specimens are drawn from naked eye observation and some fanciful, others like *Festina Lente*, plate 2 in part 1 of the book, *Archetypa studiaque patrisä, Una Hirundo non Faciotuer*, plate 6 and *Mors ulitma Linea Rerum,* plate 5 in part 2 indicate the use of a magnifying glass, for example the eyes on flies and the detail on smaller insects like mosquitoes. The scale these are drawn in relation to plants and animals on the same plate also suggests a lens of some description has been used. Similarly the naturalist Thomas Mouffet has used simple lenses for his contribution towards the 'Theatre of Insects' prepared in manuscript form in 1590 but not published until 1634. Mouffet has some of his work in Conrad Gesner's unpublished book, *Theatrum Insectorum*, Royal College of Physicians of 1588-1604.[65]

---

[65] See Ford, W., 'Development of our early knowledge concerning magnification', *Science*, Vol. 79, No. 2061, June 20 1934, pp. 578-581 and http://www-gdz.sub.uni-goettingen.de/cgi-bin/digbib.cgi?PPN371060702

Edward Wotton, Gesner, Thomas Penny and Moffett, the *Authores ex quibus hoc Opus* to *Insectorvm Sive Minimorum Animalivm Theatrvm*, of 1634 reads like a who's who of the world to that point. The hall is packed to the rafters. I manage to catch up with Gesner as he is chatting to his fellow authors, 'Gesner' I marvel, 'You have achieved published work on an encyclopaedic scale.' 'Oh' he replies humbly, 'my achievements have mostly been possible because of the open and honest correspondence I established with the leading naturalists throughout Europe who, in addition to their ideas have been generous enough to send me plants, animals and gems.' 'But you are living in a time of extreme and personally dangerous religious tension.' 'That is so however it is important for my work that I maintain friendships on both sides of the recently activated Catholic-Protestant chasm'. Exceedingly well read, Gesner is even attempting to establish a 'universal library' of all books in existence. He hands me his list of fossils while continuing his previous conversation. For Gesner a fossil is any interesting object found in the ground, including items of inorganic origin. He divides the fossils into 15 categories, and accustomed as I am to tables composed during the Enlightenment period, his list has smile value,

1. Those whose forms are based upon, have some relation to, or suggest the geometrical conception of points, lines or angles

2. Those which resemble or derive their name from some heavenly body or from one of the Aristotelian elements

3. Those which take their name from something in the sky

4. Those which are named after inanimate terrestrial objects

5. Those which bear a resemblance to certain artificial things

6. Things made artificially out of metals, stones, or gems

7. Those which resemble plants or herbs

8. Those which have the form of shrubs

9. Those which resemble trees or portions of trees

10. Corals

11. Other sea plants which have a stony nature

12. Those which have some resemblance to men or to four-footed animals, or are found within these

13. Stones which derive their names from birds

14. Those which have a resemblance to things which live in the sea

15. Those which resemble insects or serpents

What is different to what others have done before him is the use of fine woodcut illustrations to accompany his text. He acknowledges the help of Lucas Schan of Strasbourg in executing these. It is possible he had the help of other artists and may have even done a few himself. If I have enough money I can afford to have these monochrome prints hand coloured. Earlier illustrations were engravings or hand drawn but that took too much time to produce. Gesner too is using a magnifying lens for finer detailed plants and animals. In 1556, he published a pamphlet entitled *Little Commentary on Rare and Admirable Plants Called Lunariae, Either Because They Glow at Night or for Some Other Reason.*[66] I am still smiling as I search out Federico Cesi.

In 1603 Cesi, along with four of his friends, founded the first Scientific Academy in Europe, the Accademia dei Lincie, which in 1611 included Galileo as a member. This first institution is a turning point in the way science is done. In 1610 Galileo published *Sidereus nuncius* in which he describes not only how he heard of the telescope but crucially made the announcement that Jupiter had four satellites. Jupiter can be seen with the naked eye but to see the moons requires a telescope. This is the first

---

[66] Most of the preceding on Gesner is at http://www.strangescience.net/gesner.htm

recording of distant objects that cannot be seen without lens help.[67]

In 1625 and 1630 Francesco Stelluti recorded the first observation from nature under a microscope. In 1630 he published his findings, including drawings of a bee and a weevil perfected five years earlier in collaboration with Cesi, in his Italian translation of the Latin satires of Persius. The microscope they used was probably supplied by Galileo who had learned of the techniques for making microscopes and telescopes from the Dutch spectacle maker Hans Janssen or his son Zacharias or Hans Lippershey. There is no evidence to support the case that any of these men discovered the idea of putting one lens in front of another - a surprising oversight given side-by-side lenses had been used as reading glasses in Italy since 1280. Reports that Marco Polo observed Chinese wearing spectacles doesn't appear in his account, although the Chinese did grind lenses and wear sunglasses, but not as an aid to vision. The Chinese themselves believe spectacles were invented in Arabia in the 11th century. My concern here is not so much with

[67] Bardell, D., 'The First Record of Microscopic Observations', *BioScience*, Vol.33 No. 1, January 1983, pp. 36-38.

origins but rather the impact this invention and the telescope had on seeing the social.[68]

Gassendi, the very busy and well-rounded scholar who nudged me on the arm earlier, is still here. He is a professor of philosophy, a priest, astronomer and professor of mathematics at the Royal College of France. He is a friend of Mersenne, an assistant of Peiresc, an avid supporter of Galileo and a rival of Descartes. It is his role as biographer that most interests me at the moment. His subjects include Epicurus, Copernicus, Brahe and Peiresc. His biography of Peiresc was written in 1641 and translated into English 16 years later. I am skimming the text to find evidence that Peiresc knew of and used lenses. The first complete mention is in a letter he received from Pignorius informing him of the discovery in 1610 of four moons about Jupiter by Galileo using the newly invented

---

[68] An incomplete pair of rivet spectacles discovered under the floorboards of a German convent have been dated to circa 1400. In 2000 a complete pair, probably of very early date, was excavated in the Dutch town of Bergen-op-Zoom. This pair has now been restored but is still undergoing scientific evaluation. A fifteenth century pair of spectacles (c.1440) was found near the banks of the Thames at Trig Lane in 1974 and is in the ownership of the Museum of London. The first written reference to spectacles was by Francesco Redi (1626-1697) who mentions a 1289 manuscript written by a monk who said he could 'neither read nor write without the glasses called *cochiali* for the improvement of his vision.' (Ford, W., 'Development of our early knowledge concerning magnification', *Science*, Vol. 79, No. 2061, June 20 1934, pp. 578-581.) Gassendi writes of spectacles in his life of Peiresc 'In like manner he changed somewhat of his opinion touching the Spectacles, used by pore-blind and aged people', Gassendi, P., *The Mirrour of True Nobility & Gentility*, Infinity, Haverford, USA, 2003 (1641), p.219.

telescope. Gassendi then gives an account of the early moments of lenses and mentions Jacobus Metius who was working on lenses for telescopes at the same time as Lippershey. Metius applied for a patent a few weeks after Lippershey but the Netherlands State awarded it and the contract to Lippershey. When Metius died his wish that his tools be destroyed in order to prevent anyone else from claiming invention was carried out. One instrument made by Metius had combined a microscope and telescope. Gassendi is slowly circulating the hall gathering information for his biographical works. As he pauses from a conversation to make some notes I ask him to tell me the story about early telescopes, 'Jacobus Metius of Alcmair in Holland, while he was compounding and setting together sundry sorts of glasses, to try their effects, he happened accidentally upon that same comparison and composition of a convex and a concave glasse, by which, especially the Tube being interposed, he that lookt thorow the same might see small things grow great, and things distant brought neer (whereupon the invention of the Telescopium, or Perspective Glasse is attributed to him; though Johannes Baptista Porta had already published some such thing in print) but Galilaeus only by the rumour of such a thing which he had heard, began to invent not onely the cause of the effects of the

Telescope, or Perspective-Glasse; but also the way to make one; whereupon after divers essayes and trials, he hit at last upon the way a most exact one.'[69]

So Gassendi attributes the Telescopium or Perspective-Glasse to Metius, though he does mention Johannes Baptista Porta the mathematician and inventor from Naples. Gassendi doesn't reveal the text but it could be the encyclopaedic and influential *Polygraphiae*. The invention of the camera obscura has been attributed to Porta as well though he himself made no claim. Chance and accident accompany these inventions. 'Peiresc was very excited by these inventions wasn't he' I say. 'Oh yes, with great ardency of affection, that he might obtain his Book and a Telescope, or Prospective glasse as soon as possibly he could.'[70] 'You mean Galileo's *Sidereus Nuncius*. How did he go getting both the book and telescope?' 'Though he got a Book, yet was it long ere he could obtain an exquisite Telescope, though he got some both from Italie, Holland, and Paris, as soon as they began to be made there.'[71] 'As soon as he managed to

[69] *The Life of Copernicus (1473-1543)*, Pierre Gassendi with notes by Olivier Thill, p.106.

[70] Ibid., 106.

[71] Ibid., 107.

acquire a telescope he set to studying Jupiter' 'Assuredly and he shewed the same to Varius, and other friends; and that he might lose no time he made him an Observatorie, and invited the foresaid Galterius, and kept him divers dais, and spent almost two whole years together in his observation with him. His Brother was yet at Paris: wherefore he never ceased to urge and sollicite him, till he caused divers glasses for prospectives to be made, which he sent him to the number of Fortie.'[72]

'He was also interested in the microscope' I proffer 'Yes, he it was [Ludovicus Machaltus], to whom Peireskius [Peiresc] did demonstrate in a certain beautifull Diagram, the way to multiply the species and appearance of one and the same thing, between Two Glasses, declining laterally one from another.'[73] 'I writ him word that there was a Book' he continued, 'published by Gasper Asellius Anatomist at Ticinum, wherein he shewed how he had discovered certain milky veins in the Mesentery (besides the commonly noted red ones) which probably carried Chylus; he speedily got divers of the books, which he sent up and down to Physicians, which were his friends,

---

[72] Ibid.

[73] Ibid., 152.

experimenting in Dogs, Sheep, Oxen, and most kind of animals besides, that which Asellius had written touching his rare invention.'[74] 'You informed him about the work of William Harvey, the English Physician?' 'Yes, he set out an excellent Book of the passage of blood out of the Veins, into the Arteries.'[75] 'But you disagree with Harvey because you believe there are winding passages in the heart's septum that allow this to happen' I remind him 'Yes, but I later changed my view.' 'Harvey also writes in his book' Gassendi continues 'that the inventor was Father Paul Sarpi of Venice.[76] But you asked me specifically about microscopes. Peiresc after being told of the many-pointed tongues of flies, which might be plainly seen by an Augmenting-glasse; he made many experiments in Insects of like nature, and especially in Bees, that he might thereby give occasion to Rigaltius, to mend and illustrate Pliny his Chapter of Bees.'[77]

From the beginning of its history the Royal Society has devoted much attention to the publication of

---

[74] Ibid., 177.

[75] Ibid.

[76] Ibid.

[77] Ibid., pp.177-178.

communications by its Fellows and others. Within three years from the granting of the first Charter, Henry Oldenburg, the first Secretary, began publishing *Philosophical Transactions* in March 1665. Towards the end of the introduction to the March 6 publication the author states the intentions in printing the journal,

> To the end, that such productions being clearly and truly communicated, desires after solid and useful knowledge may be further entertained, ingenious endeavours and undertakings cherished, and those, addicted to, and conversant in such matters, may be invited and encouraged to search, and find out new things, impart their knowledge to one another, and contribute what they can to the Grand design of improving Natural knowledge, and perfecting all philosophical arts, and sciences. All for the glory of God, the honour and advantage of these kingdoms, and the universal good of mankind.

It is the intention of the publishers to contribute to and perfect knowledge for a universal good. Calls were sent out to all learned people across Europe, not just Britain. The first article to mention optics was submitted to volume 3 of 1668 being an extract from the Italian

Giornale de Letterati called 'Concerning a Microscope of a new Fashion, Discovering Animals Lesser Than Any Seen Hitherto',

> Eustachio Divini hath made a *Microscope* of a new invention, wherein instead of and Eye-glass convex on both sides, there are two plano-convex Glasses, which are so placed, as to touch one another in the middle of their convex surface. This Instrument, of which *Hon. Fabri* treats largely in his *Opticks (vis. Prop. 4.6)* hath this peculiar, that it shews the Objects flat and not crooked, and although it takes in much, yet nevertheless magnifieth extraordinarily.
>
> It is almost 16 and a half inches high, and adjusted at 4 different lengths. In the *first*, which is the least, it shews lines 41 times bigger than they appear to the naked eye: In the *second*, 90 times: In the *third*, 111 times: and in the *fourth* 143 times. Whence one may easily calculate, how much it augments surfaces and solidities.[78]

---

[78] *Philosophical Transactions*, Vol.3, 1668, p.842.

For an example of the power of this microscope the animal observed is the same size as a grain of sand,

> As they viewed with this Microscope the little grains of sand searched, they perceived an Animal with many feet, its back white and scaly, but less than any of those hitherto observed. For, although the Microscope shewed every grain of sand as big as an ordinary Nut, yet this Animal appeared no bigger than one of those grains of sand seen without a Microscope. Whence may be concluded its smallness, which occasion'd one of the beholders to give it the name of the *Atome of Animals*.[79]

Leeuwenhoek is gathering all manner of small animals and items to take home and view, then report back to the donors, through *Philosophical Transactions*. He isn't demonstrating his latest microscopes here but has recently submitted a letter, via Dr Regnerus de Graaf giving some of his results and indicating his microscopes are exceeding the work done by Eustachio Divini.

'I note you have done some detailed observation on bees Leeuwenhoek, including the sting and what you call the

---

[79] Ibid.

Scrapers, Arms and Wipers that they use when gathering wax and honey-substance from plants,' 'Ah yes. I have also studied closely the eye of a Bee.' 'What did you find?' 'I took the eye out of the head, exposing its innermost part to the Microscope; I find that the Bee receives her light Just with the same shadow as we see the Honeycombs: Whence I am prone to collect, that the Bee works not by art or knowledge, but only after the pattern of the light received in the Eye.'[80] 'So you believe the bee isn't working from intelligence but just copies refracted light patterns to make honeycomb.' 'Yes, I also saw how a Lowse draws up food, the skin on its head and the two Claws, one foot is the structure of that of an Eagle's, but the other of the same foot stands out straight, and is very small; and between these two claws there is a raised part or knob, the better to clasp and hold fast the hair.'[81] It is clear from what he is telling me the microscopes he is using are significantly more powerful than his contemporaries. But I know he is cagey about how he achieves his results. I ask him to show me how to use one of his microscopes. I look through the single lens but can't see any detail. I shift my position to allow more

---

[80] *Philosophical Transactions*, April 28, 1673, p.6038.

[81] Ibid.

sunlight onto the specimen. This helps slightly but if Leeuwenhoek hadn't told me it was a louse mounted on the pin I wouldn't know what I am looking at and I certainly couldn't see the detail he describes. I tell him the difficulty I am having. 'It takes a lot of practice, a steady hand and excellent illumination' he explains and adds, 'Also I use a new lens for each object to be viewed. I don't have any more with me; they are at home in the Netherlands'. His tone changes and he reminds me, in case I had forgotten, he does not 'gladly suffer contradiction or censure from others.'[82] 'You will find it helps if you put this dark ground or black background behind the specimen and let the light in from the side.'[83] 'When there is no sunlight I use a candle,' he adds.[84] I try this and immediately see the difference. The louse shows up like a bright star in a clear night sky. 'This is beautiful', I tell him. He smiles and is hoping no doubt I will press him no further. But there is something more to the way he views objects, for example he reports to The Royal Society in 1677 that he has seen living animals eight times smaller than the eye of a louse in samples of

---

[82] Ford, B.J., *The Revealing Lens: Mankind and the Microscope*, p.51.

[83] Ibid., p.55.

[84] *Philosophical Transactions*, Vol. 12, 1677, p.824.

water. In a now famous letter for publication in *Philosophical Transactions* he explains, in some 18 pages, how he has done detailed and prolonged studies of microscopic living organisms. He has excelled himself as a discoverer of a whole new universe beneath the lens.[85]

Leeuwenhoek is excited and enthusiastic about his discoveries. In fact he seems more willing to share his findings and methods this time than on previous occasions. I decide to ask him again about how he achieves his remarkable results. He is instantly guarded 'I intend to keep my method of observing the tiniest creatures to myself. I have shown you how I observe under the microscope. You will need to practice some more and keep your unsteady hand still. That is all I am going to say on the subject.'[86] He collects up his microscopes and attachments and samples and exits the hall. I sit down for a few minutes speculation on what I have just seen and heard and to think about Leeuwenhoek's microscopes. He tells me he uses a single lens but compound lenses were in use in the 1660's. Leeuwenhoek is quick to take up ideas from people like Swammerdam, Hooke, Divini, de Graaf and Huygens,

---

[85] Ford, B.J., *The Revealing Lens: Mankind and the Microscope*, p.53.

[86] Ibid., p.55.

and he does not always acknowledge his debt, like the influence reading Hooke's *Micrographia* on a visit to London in 1668 had on spurring his early interest in the microscope. Without a compound microscope he would not have been able to observe bacteria. The glass lenses he makes are of very high quality (clarity and minimal distortion are a feature) and he knows the importance of bright illumination and using a dark ground, but this isn't enough for objects smaller than insect heads and eyes and plant and animal fragments. Leeuwenhoek makes no secret of the fact that he has a secret. A secret he won't share with his closest friends. A secret he is prepared to take to his grave. He does leave a hint however, in a letter he wrote to the Royal Society in 1717, explaining his hands are too weak to continue his observations. He duly recovered from this disability and continued studying for another five years, however on his death the 172 lenses mounted between plates - with handmade mounts - that were sold at auction, along with 247 microscopes, are most likely to hold the key to his secret. What is missing from this auction is also part of the secret. When the microscopes and lenses were gathered for auction the mounted subjects that remained were conventional subjects like insect parts and minerals. There were no specimens of smaller subjects like bacteria and protozoa.

It is highly likely that Leeuwenhoek was using hand held lenses to augment the fixed lenses on his microscopes, thus replicating the precision of a compound microscope but with greater power of magnification. Having a steady hand is crucial to observation of this kind.[87]

---

[87] For a detailed explanation see Ibid., pp.59-60.

CHAPTER THREE

Camera Obscura

I remember an earlier conversation with Alberti where he said 'more could be said about these reflections, which relate to those miracles of painting which many of my friends have seen made by me previously in Rome.'[88] The 'miracles' he was talking about are described more fully in the anonymous *Vita* in *opera volgari* 'the pictures, which were contained in a very small box, were seen through a tiny aperture. There you were able to see very high mountains and broad landscapes around a wide bay of sea, and, furthermore, regions removed very distantly from sight, so remote as not to be clearly seen by the

---

[88] Alberti, L.B., *On Painting*, p.97, note 7. A useful guide to the camera obscura is *Photographic Literature 1960-1970*, Albert Boni (Editor), A bibliographic guide, Morgan and Morgan Inc. Publishers, New York, 1972(1962), Bibliotheque Nationale. See also mediaeval collections on astronomy and optics, particularly Arabic texts held at the Bibliotheque Nationale and at the Science Museum London. Also Svetlana Alpers, *The Art of Describing,* Chicago: University of Chicago Press, 1983, while not agreeing with her strong position, a useful resource.

viewer. He called these things 'demonstrations'...he called one of them 'daytime' and the other 'nighttime.'[89]

Daniel Barbaro's *La Pratica della Perspectiva, Opera molto utile a pittori, scultore ed architetti*,[90] contains the first description of a camera obscura fitted with a lens and the first description of an adjustable diaphragm to sharpen the image. I ask him about its use. Like da Vinci he doesn't hesitate 'I encourage, unapologetically, the use of the camera obscura for artistic endeavours. I know at this time my adapted technique for using the camera obscura is being replaced by a simple box rather than an entire room, as in my demonstrations, however my treatise is still important'. 'Ah yes, I understand your treatise on perspective was extremely influential during the sixteenth century and the improvement in image definition brought about by the inclusion of a lens and adjustable diaphragm made it useful for tracing.' 'Yes', he replies and proceeds to tell me how it works. 'Close all shutters and doors until no light enters the camera except through the lens, and opposite hold a piece of

---

[89] Ibid., p.98. For a longer analysis see 'The Authorship of the Vita Anonyma of Leon Battista Alberti,' Renée Watkins, *Studies in the Renaissance* Vol. 4 (1957), pp. 101-112. Bonucci 1843-1849 copied from *Vita anonyma* of Lorenzo Mehus editor from 1751, held by some to be autobiographical notes of Alberti or friends of.
[90] Folio, Berominier, Venice 1568.

paper, which you move forward and backward until the scene appears in the sharpest detail. There on the paper you will see the whole view as it really is, with its distances, its colours and shadows and motion, the clouds, the water twinkling, the birds flying. By holding the paper steady you can trace the whole perspective with a pen, shade it and delicately colour it from nature.'[91] 'And the type of lens used is important?' I ask. 'You should choose the glass which does the best, and you should cover it so much that you leave a little in the middle clear and open and you will see a still brighter affect. The glass that works best is a bi-convex lens taken from a pair of ordinary reading spectacles. I have also experimented with concave lenses but so far I haven't had much success.'[92]

In 1604 Kepler published *Paralipomena ad Vitellionem*. He took his ideas on optics from Francesco Maurolico who had written that every point of a body emits rectilinear rays in all directions, rays that are neither visual nor solar but simply geometrical. Maurolico's *Photismi de lumine et umbra* concerns the refraction of

[91] micro.magnet.fsu.edu/optics/timeline/people/barbaro

[92] Ibid. Barbaro has been credited with translating Vitruvius' ten books on architecture.

light and attempted to explain the natural phenomenon of the rainbow. It was completed in 1521 but was published posthumously in 1611. Maurolico also studied the camera obscura. *Somnium* was written by Kepler between 1593 and 1630. He began this work as a student dissertation (though never presented) in which he defends the Copernican doctrine of the motion of the Earth, suggesting that an observer on the Moon would find the planet's movements as clearly visible as the Moon's activity is to the Earth's inhabitants. In 1609 he added the dream framework, and after another 13 years drafted a series of explanatory notes reflecting upon his turbulent career and the stages of his intellectual development. The book, edited by his heirs, including his son Ludwig Kepler and Jacob Bartsch after his death in 1630, was published posthumously in 1634. *Somnium* presents a heliocentric world-picture, based on the observations of Copernicus, and inverts the meaning and place of observation in the hierarchy of knowledge to the top. Kepler manages this by critically adapting Lucian's *True Story* and Plutarch's *The Face on the Moon* along with the help of Maurolico's optics. Astronomical knowledge is justified while scientific observation offers a way out of his dream. He begins in a light-hearted, ironic and jocular manner but soon becomes quite serious.

Tellingly Kepler - accompanied by his mother - begins his moon journey in a camera obscura. It proved to be a dangerous journey for his aged mother.

'My mother went away from me to the nearest crossroads. Raising a shout, she pronounced just a few words in which she couched her request. Having completed the ceremonies, she returned. With the outstretched palm of her right hand she commanded silence, and sat down beside me. Hardly had we covered our heads with our clothing (in accordance with our covenant) when the rasping of an indistinct and unclear voice became audible.'[93] In his notes Kepler informs us that his mother performed a magic ceremony which is in accord with the teachings of astronomy where 'every prompt answer requires repose, recollection of ideas, and set words,' then proceeds to describe his demonstrations of a camera obscura. 'During those years in Prague I often carried out a special procedure in connection with a certain observation. Whenever men or women came together to watch me, first, while they were engaged in conversation, I used to hide myself from them in a nearby corner of the

---

[93] *Kepler's Somnium, The dream, or Posthumous Work on Lunar Astronomy*, Translated, with a Commentary by Edward Rosen, Dover Publications, New York, 2003 (1967), pp.14-15.

house, which had been chosen for this demonstration. I cut out the daylight, constructed a tiny window out of a very small opening, and hung a white sheet on the wall. Having finished these preparations, I called in the spectators. These were my ceremonies, these were my rites. Do you want characters too? In capital letters I wrote with chalk on a black board what I thought suited the spectators. The shape of the letters was backwards (behold the magical rite), as Hebrew is written. I hung this board with the letters upside down in the open air outside in the sunshine. As a result what I had written was projected right side up on the white wall within. If a breeze disturbed the board outside, the letters inside wiggled to and fro on the wall in an irregular motion.[94] The crossroads represents both the point of image reversal demonstrated in his house and the astronomical crossroads in the heavenly patterns, based on the observations of Brahe. Kepler enjoyed the playful aspects of his demonstrations, like a showman. His note 49 illustrates what 'Covering our heads with our clothing' means; 'With this very rite (ha, how magically magical!), shortly before I conceived the plan of this book, we had observed a solar eclipse on October 12 1605. You

---

[94] Ibid., note 44,46 and 47 on p.57.

remember, O envoys from the Count Palatine of Neuburg, because you were present. For on the balcony of the pavilion in the emperor's gardens we lacked a dark room. Therefore we covered our heads with our coats and kept out the daylight in that way.'[95] When the travellers reach the moon (Levania) and disembark their ship and go ashore they 'quickly withdraw into caves and dark places.'[96] As Kepler explains in his notes the allegory,

> compares the journey through the observation of eclipses; the sun, to political business; the dark caves of the moon, to seclusion and scholastic obscurity; the time spent in the caves, to continuous speculation based on observations of the eclipses. In Prague I had a residence in which no spot was more suitable for observing the sun's diameter than the underground beer cellar. From the floor of the cellar I used to aim an astronomical tube, described in my *Optics*, through an opening at the top toward the noonday sun on the days of the solstices.[97]

---

[95] Ibid., p.58.

[96] Ibid., p.16.

[97] Ibid., p.75.

Unfortunately his developing note on the allegory is missing. His astronomical tube is not a telescope and doesn't contain lenses. Solar eclipses had long been observed by formal and makeshift camera obscura's. At the end of the dream Kepler returns to earth from the moon and to himself, right way round and 'found my head really covered with the pillow and my body with blankets,'[98] thus returning to the camera obscura. The view from the moon allows Kepler to invert the current belief that the static earth is the centre of the universe and, following Copernicus, shows it revolving on its own axis and around the sun.[99]

Jean-Francois Niceron touchingly wrote in the Paris of 1646 that 'si Deus faverit, otiumque et vires ex eius immensa bonitate suppenditent' (if God will grant me, by his immense goodness, the leisure and the strength), he will publish a monograph of all the aspects of his studies in geometric optics. God didn't grant him the time and his friend and mathematics mentor Mersenne only

---

[98] Ibid., p.29.

[99] See also 'Shadows of Instruction: Optics and Classical Authorities in Kepler's "Somnium"', Raz Chen-Morris, *Journal of the History of Ideas*, Vol. 66, No. 2. (Apr., 2005), and 'What is the History of Medieval Optics Really About?' Smith, M.A.

managed to include the first two of the planned four books in *Thaumaturgus Opticus seu Admiranda Optices, per Radium Directum.* Included is an introduction based on methods for drawing the five regular solids on two dimensional surfaces, especially on the curves and arcs often seen on church ceilings, while the second part deals with optics, or direct vision. What were not included were the third book on catoptrics, or viewing by reflections in flat, cylindrical and conical mirrors and the fourth on dioptrics, or viewing by refraction through lenses. Despite these omissions Niceron has left us some beautiful illustrations of the inverted images formed on a wall or screen in a darkened room camera obscura. He also left us a shorter version of *Thaumaturgus* titled *La perspective curieuse, ou magie artificielle des effets merveilleux* (Curious perspective, or artificial magic of marvelous effects), published in French in 1638.[100] The etched frontpiece portrait to this treatise has Niceron carefully demonstrating perspectival drawing including shadows on a wall and the floor, indicating two light sources. He is dressed in Saint Francois de Paule monk's habit. Niceron acknowledges Alberti and Durer and Barbaro in the introduction as well as mentioning a large

---

[100] *La perspective curieuse, ou magie artificielle des effets merveilleux*, November 5, 1651 F. Hilarion De Coste and Frere Ambroise Granjon publication. Available online.

array of scientific instruments, including astronomical aids. The illustrations show a variety of anamorphic subjects, perspective, parallelograms, the use of drawing instruments like pantographs and rulers, lenses like a kaleidoscope for tricks of the eye, mirrors, ground crystals, room sized camera obscura's (drawn to scale with tracing demonstration), the drawing of complex and simple geometrical shapes and objects in two dimension and objects occupying space. There is a feeling of playfulness about this book, though its subject is treated carefully and is mathematically exact. Niceron believed in the Copernican theory of the solar system and revered the work of Galileo, in particular his observations with the aid of a telescope. Mersenne first learned of Galileo's death in a letter sent to him from Rome by Niceron, with its moving statement that 'mathematicians must now mourn because their glory has been extinguished with the death of Galileo.'[101]

In November 1614 Christopher Scheiner was called to Innsbruck by Maximilian III to discuss astronomical and mathematical questions and to help him sort out his

---

[101] Gould, S.J and Shearer, R.R, 'Drawing the Maxim from the Minim: The Unrecognized Source of Niceron's Influence Upon Duchamp', *tout-fait*, Issue 3, 2000. Available online www.toutfait.com

telescope. Maximilian had been given an astronomical telescope with two convex lenses. Unfortunately the images were upside down and wrong way round. Scheiner added a third lens that allowed Maximilian to see the countryside while standing upright. Around this time a portable camera obscura was developed by Scheiner as was a walkable camera obscura. On the death of Maximilian, Leopold V was appointed imperial representative of Tyrol and of the Upper Provinces and he also put his trust in Scheiner. Scheiner's *Oculus hoc est: Fundamentum opticum* was published in Innsbruck in 1619. *Oculus* is organised into three parts. The first part treats the anatomy of the eye, the second the refraction of the light ray inside the eye and the third deals with the retina and the visual angle. Like Kepler before him he found that the retina is the seat of vision and that the optic nerve transmits the images from the retina to the brain. Scheiner's detailed study of the eye predates Descartes work and he was particularly impressed by the optical analogy of the eye and the camera obscura. Vigorous debate about the nature of seeing marked the middle decades of the 1600's. There is evidence suggesting that a number of unpublished letters and manuscript treatises on optics and vision circulated privately between Peiresc, Gassendi, Mersenne, Scheiner,

Liceti, and Boulliau during the 1630s, and thereafter, members of this group joined debates between Descartes, Fermat, Hobbes, and Mydorge.[102] The challenge for these people was to combine a coherent geometry of sight with a physical explanation of how light made vision possible. Scheiner also published a book on pantographs, which he invented, called *Pantographice seu ars delineandi* in Rome in 1631. The inventory of Leopold's library contains works by Brahe and Galileo - and Leopold maintained a friendly correspondence with Galileo. On May 23, 1618, Leopold received from Galileo telescopes and his treaty on sunspots, the *Discorso del Flusso e Reflusso del Mare*.

'But language, as often as not, and even more often than not, is deceptive' says Alexandre Koyre. 'It is made by and for common use; it is based on images. It is not the embodiment of genuine thought, of a thought clearly conscious of its own requirements. This thought, Cartesian thinking, starts with the infinite, the perfect. It conceives infinite space before it inscribes figures in it. It conceives God before proceeding to understand man.'[103] A paragraph from a letter Descartes wrote to Pierre

---

[102]   web.clas.ufl.edu/users/rhatch/pages/03-Sci-Rev/SCI-REV-Home/resource-ref-read/vision/08sr-vision.htm

[103] Alexandre Koyre, pp.xxxvii-xxxix introduction to *Descartes Philosophical Writings*, Nelson, 1964.

Chanut in 1647 tells us something about his belief in the infinite, 'The six days of the creation are indeed described in Genesis as though man were the principal object of creation; but one could say that since the account in Genesis was written for man, the Holy Ghost saw fit to give particulars principally of what concerns man, and that indeed nothing is mentioned there except in its relation to man.'[104] Descartes argument for the existence of God is an ontological one, taken in main from scholastic philosophy. It depends on the distinction between existence and essence, with God representing the perfect Being who necessarily exists.

In his third *Meditation* Descartes begins 'I will now shut my eyes, stop my ears, withdraw all my senses; I will even blot out the images of corporeal objects from my consciousness; or at least (since this is barely possible) I will ignore them as vain illusions. I will discourse with myself alone and look more deeply into myself; I will try to grow by degrees better acquainted and more familiar with myself.'[105] He is keeping the images, along with his vanity, despite his silent protests. In Discourse V of

---

[104] *Descartes Philosophical Writings*, p.295.

[105] Ibid., p.76.

Dioptrics he begins 'You see, then, that sensation does not require that the soul should contemplate any images resembling the objects of sensation. For all that, the objects we look at do in fact produce very perfect images in the back of the eyes. This has been explained by a most ingenious comparison.'[106] Descartes is making it very clear by the use of the following analogy that human vision, thought, and the workings of the camera obscura, are the same. The emptied out interior space of the mind resembles the darkened empty room 'If a room is quite shut up apart from a single hole, and a glass lens is put in front of the hole, and behind that, some distance away, a white cloth, then the light coming from external objects forms images on the cloth.'[107] And he continues 'Now it is said that this room represents the eye; the hole, the pupil; the lens, the crystalline humour – or rather, all the refracting parts of the eye; and the cloth, the lining membrane, composed of optic nerve endings.'[108] Descartes then describes a demonstration, following Kepler, where the eye of a newly dead man or an ox or other large animal is substituted for the lens in the hole

---

[106] Ibid., pp.244-245.

[107] Ibid., p.245.

[108] Ibid.

in the camera obscura.[109] The image produced 'represents in natural perspective all the objects outside.'[110] The white cloth is a *tabula rasa* analogous with his ignoring of the vain illusions of the corporeal images in his mind, making available to the intellect at each viewed instant, a surprisingly true representation of the outside world. But of course the images are attractive and remain even if we close our eyes, as do bright light after-images. Crucially however, Descartes, at the very centre of his enterprise, replaces human uncertainty and the vagaries of human vision with an objective – disembodied - view of the world, represented by the mechanical camera obscura.[111]

As early as November 10 1619 Descartes was thinking in this emptied out space 'Just as comedians are counselled not to let shame appear on their foreheads, and so put on a mask: so likewise, now that I am about to mount the stage of the world, where I have so far been a spectator, I

---

[109] One adaptation to the lens of the camera obscura was the scioptric ball or 'ox-eye' which was a wooden sphere fitted with two lenses and attached to the wall. There are two on display in the Science Museum London.

[110] *Descartes Philosophical Writings*, p.245.

[111] See Crary, J., *Techniques of the Observer*, MIT Press, Cambridge, 1992.

come forward in a mask.'[112] But doing so he abandoned action – a military career in the army of Maurice of Nassau – for contemplation, beginning in the stove heated chamber in which he sat alone during that winter, or at least all day, and emerged – so the story goes - with his philosophy half finished. Descartes then looks out from his enclosed room (analogous with the camera obscura) to describe his explanation for thought, 'I chanced, however, to look out of the window, and see men walking in the street; I say in ordinary language that I 'see' them, just as I 'see' the wax [extracted honeycomb wax]; but what can I 'see' besides hats and coats, which may cover automata? I judge that they are men; and similarly, the objects that I thought I saw with my eyes, I really comprehend only by my mental power of judgement.'[113] At the beginning of his *Treatise on Man* Descartes writes 'These men will be composed, as we are, of a soul and a body. First I must describe the body on its own; then the soul, again on its own; and finally I must show how these two natures would be joined and united in order to constitute men who resemble us.'[114] The

---

[112] *Descartes Philosophical Writings*, p.3.

[113] Ibid., pp.73-74.

[114] *The Philosophical Writings of Descartes Vol.1*, Cambridge University Press, Cambridge, 1985, p.99.

translators note that by 'these men' Descartes means the 'fictional men he introduced in an earlier (lost) part of the work. Their description is intended to cast light on the nature of real men.'[115] Although it is not clear that Descartes is employing a heuristic device or if he is carefully covering his thoughts out of concern for the church, his idea has been interpreted as referring to 'real men' in the world. What is important is that Descartes separates the soul from the body and presents two parallel but independent worlds, that of mind and matter, both of which can be studied independently. An easy extension made by later scholars was that not only are animals automata, following Descartes, so are men, thus making consistent the materialist argument.

Louis De La Forge was a medical practitioner. He is mostly remembered because he propagated the ideas and doctrines of Descartes to a larger audience. His works, among which is the commentary and medical illustrations to the first French edition of *Traité de l'homme*, contributed largely to the spread of Cartesianism. His knowledge of medical science made it possible for him to apply some of Descartes' statements

---

[115] Ibid.

to anatomy and physiology. Further, he published a *Traité de l'Esprit* in an occasionalistic vein, of which philosophical trend some consider him the founder. De La Forge wrote to *Philosophical Transactions* in 1668 informing them that Descartes writings are now available in Latin and that other letters were being translated by him and further he would include his own tracts on 'Man and the union of the rational soul with the body' shortly.[116] Occasionalism denotes the theory of occasional causes. It is 'not the body that gives rise to perception, nor the mind that causes the motion of the limbs which it has determined upon - neither the one nor the other can receive influence from its fellow or exercise influence upon it; but it is God who, 'on the occasion' of the physical motion (of the air and nerves); produces the sensation (of sound), and, 'at the instance' of the determination of the will, produces the movement of the arms. The systematic development and marked influence of this theory was more or less clearly announced by the Cartesians Cordemoy and De la Forge.'[117]

The celebrated Mr Hook ushers me into the darkened room and with the pitch of a showman begins his banter

---

[116] *Philosophical Transactions,* 1668, p.810.

[117] www.humanitiesweb.org/human.php

'This is a new idea. It is both easy and obvious and produces effects onely very delightful, but to such as know not the contrivance, very wonderful.' I indicate some knowledge of the camera obscura so he follows with 'Spectators not well versed in *Opticks* that should see the various Apparitions and Disappearances, the Motions, Changes, and Actions, that may this way be represented, would readily believe them to be super-natural and miraculous, and would easily be affected with all those passions of Love, Fear, Reverence, Honour, and Astonishment, that are the natural consequences of such belief.' Not needing encouragement he is into his stride, 'had the *Heathen* Priests of old been acquainted with it, their Oracles and Temples would have been much more famous for Miracles of their Imaginary Deities. For by such an Art as this, what could they not have represented in their Temples?' Apparitions of Angles, or Devils, Inscriptions and Oracles on Walls; the Prospect of Countryes, Cities, Houses, Navies, Armies; the Actions and Motions of Men, Beasts, Birds, &c, the vanishing of them in a cloud, and their appearing no more after the cloud is vanished: And indeed almost any thing, that may be seen, may by this contrivance be vividly and distinctly represented, in such a manner, that, unless to very curious and sagacious persons, the means how such Apparitions are made, shall not be discoverable.'[118]

Mr Hook settles down and adopts a more scientific tone as he explains his process to me, 'Opposite to the place or wall, where the Apparition is to be, let a Hole be made of about a foot in diameter, or bigger; if there be a high Window, that hath a Casement in it, 'twill be so much the better. Without this hole, or Casement open'd at a convenient distance, (that it may not be perceived by the Company in the room) place the Picture or Object, which you will represent, inverted, and by means of Looking-glasses placed behind, if the picture be *transparent,* reflect the rayes of the Sun so, as that they may pass through it towards this place, where it is to be represented.' He then goes into fine detail, describing the various lenses, both reflecting and refracting, the apparatus, the operator's, the room and the entire operation. He also promises to describe how he makes landscapes both during the day in a lit room, the use of sun-beams and how, at night candles, torches, lamps and other bright lights can be placed around objects to make them appear on the wall. Mr Hook does not content himself with 'bare speculation, but put the same in

[118] *Philosophical Transactions,* Monday August 17 1668, pp.741-742.

practice some years since, in the presence of several members of the Royal Society, among whom the *Publisher* had the good fortune to see the successful performance of what is here answered.'[119]

A showman to be sure but I have just read a letter Constantijn Huygens (who began a lifelong friendship and correspondence with Descartes in 1630) wrote home in 1662 where he describes the abilities of a camera obscura, 'It is impossible to express its beauty in words. The art of painting is dead, for this is life itself, or something higher, if we could find a word for it.'[120]

---

[119] Ibid., p.743.

[120] Wheelock, A.K., 'Constantijn Huygens and Early Attitudes to the Camera Obscura,' *History of Photography 1*, 1977, pp. 93-103.

CHAPTER FOUR

The *Encyclopédie*

The Frontpiece to the 1751 edition of the *Encyclopédie* was drawn by Charles-Nicolas Cochin and engraved by Bonaventure-Louis Prévost. The accompanying caption explains the illustration;

> Beneath a temple of Ionic architecture, sanctuary of Truth, we see Truth wrapped in a veil, radiant with a light which parts the clouds and disperses them. On the right of Truth, Reason and Philosophy are engaged, the one lifting the veil from Truth, the other in pulling it away. At her feet Theology, on her knees, receives her light from on high. Following the line of figures, we see grouped on the same side Memory, and Ancient and Modern History; History is writing the annals, and Time serves as a support for her. Grouped below are Geometry, Astronomy, and Physics. The figures below this group show Optics, Botany, Chemistry, and Agriculture. At the bottom are several Arts and

Professions that proceed from the Sciences. On the left of Truth we see Imagination, who is preparing to adorn and crown Truth. Beneath Imagination, the Artist has placed the different genres of Poetry – Epic, Dramatic, Satiric, and Pastoral. Next come the other Arts of Imitation – Music, Painting, Sculpture, and Architecture.

These two opening elements set the tone for the *Encyclopédie* project that was to last until 1780 by which time the prime mover and editor, Diderot, and the author of the *Preliminary Discourse* and the sections on mathematics and physics, amongst other things, d'Alembert, were no longer involved. The *Encyclopédie* is a visibility technology of the highest order.

'In a word, Madame' wrote d'Alembert to Madame du Deffand on December 22 1752 about his *Preliminary Discourse* to the *Encyclopédie*, 'I can assure you that while writing this work, I had posterity before my eyes at every line.'[121] Montesquieu, like many of his contemporaries, was extremely complimentary 'You have given me great pleasure. I have read and reread your

---

[121] d'Alembert, J.L.R., *Preliminary Discourse to the Encyclopedia of Diderot*, The University of Chicago Press, Chicago and London, 1995(1751), p.ix of the introduction.

Preliminary Discourse. It has strength, it has charm, it has precision; richer in thoughts than words, likewise rich in sentiment – and my praises might go on.'[122] Diderot, for sound reason, has chosen d'Alembert to be the spokesperson for the *Encyclopédie*, a role he has taken on with passion, so I go directly to him to pose my questions. 'You are very straightforward about what you and your 'society of men of letters' are doing in the *Encyclopédie*.' He smiles instantly, 'Yes, the work whose first volume we are presenting today has two aims. As an *Encyclopedia*, it is to set forth as well as possible the order and connection of the parts of human knowledge. As a *Reasoned Dictionary of the Sciences, Arts, and Trades*, it is to contain the general principles that form the basis of each science and each art, liberal or mechanical, and the most essential facts that make up the body and substance of each.'[123] 'When you say science, what do you mean and what is included?' I ask. 'I take direction from my energetic friend Diderot on this

---

[122] Ibid.

[123] d'Alembert, J.L.R., *Preliminary Discourse to the Encyclopedia of Diderot*, p.4. For the *Encyclopédie* proper d'Alembert has written articles on Fenêtre, Géométrique, Géométriquement, Scenographie: Perspective, Anamorphose: Perspective et Peint, Caricature, peint, Copernic, and École, Beaux-Arts, as well as the articles on Etoile, fortific, Defait: Vaincu: Batu, art milit. and Dictionare de Sciens et d'Arts, Tant Libéraux que Méchaniques, as well as articles as diverse as Genève, Formulaire and Ecclésiastique.

question. For him the encyclopaedic order might be created either by relating our different kinds of knowledge to the various faculties of the soul (this is the system we have used), or by relating them to the entities they take as their object; and this object may be one of pure curiosity, or a luxury, or a necessity. Science in general may be divided into science of things and of signs, or into concrete or abstract sciences. The two most general causes, art and nature, also yield an elegant and broad distribution. Others will found in the distinction between the physical and the moral, the extant and the possible, the material and the spiritual, the real and the intelligible. Does not all we know derive from the use of our senses and our reason? Is it not either natural or revealed? Is it not either words, or things, or facts? It is therefore impossible to banish arbitrariness from this broad primary distribution.'[124] 'So any human endeavour that is observed, contemplated from different approaches, has technically collected rules which are then disseminated, is a science, including theology, history and philosophy, as well as physics and mathematics?' I summarise. 'Yes,' is his quick response. The task appears overwhelming but d'Alembert assures

---

[124] *The Encyclopedia of Diderot & d'Alembert Collaborative Translation Project*, University of Michigan Library, article online at http://name.umdl.umich.edu/

me they are going to deal with each part 'one after another' as well as giving an account of the means by which they have tried to satisfy the double objective of order and connection.

'You seem to have a very practical outlook on what the task entails and how you are going to achieve it,' I advance. 'If one reflects somewhat upon the connection that discoveries have with one another, it is readily apparent that the sciences and the arts are mutually supporting, and that consequently there is a chain that binds them together. But, if it is often difficult to reduce each particular science or art to a small number of rules or general notions, it is no less difficult to encompass the infinitely varied branches of human knowledge in a truly unified system.'[125] d'Alembert, in one sentence, has placed the rationalist foundation stone to the entire edifice of the *Encyclopédie* project. Even though he expresses the difficulty there is a confidence about this man that his 'society of men of letters' can reduce knowledge down by reason alone and produce, within the covers of the *Encyclopédie*, a unified system of knowledge for all human endeavour. I am starting to get a sense of the influence of Epicurus here so press d'Alembert

---

[125] d'Alembert, J.L.R., *Preliminary Discourse to the Encyclopedia of Diderot*, p.5.

further 'How will you start?' 'The first step which lies before us in our endeavour is to examine, if we may be permitted to use this term, the genealogy and the filiation of the parts of our knowledge, the causes that brought the various branches of our knowledge into being, and the characteristics that distinguish them. In short, we must go back to the origin and generation of our ideas.'[126] The influence of Epicurus is close the surface. I tell d'Alembert about a letter Epicurus wrote to Pythocles where he said, 'you should keep in mind; for then you will escape a long way from myth, and you will be able to view in their connection the instances which are similar to these. But above all give yourself up to the study of first principles and of infinity and of kindred subjects, and further of the standards and of the feelings and of the end for which we choose between them. For to study these subjects together will easily enable you to understand the causes of the particular phenomena.'[127] 'Ah yes' he replies, 'although I only make a fleeting reference to Epicurus in my text, his work is essential to the philosophy of my close collaborator Condillac' he pauses, then continues 'I am impressed by Condillac's ideas and

---

[126] Ibid.

[127] http://www.epicurus.net/en/herodotus.html

much in his debt when writing my *Preliminary Discourse*. He and I have discussed its content on many occasions.' It is likely that Condillac relied on Gassendi's 1647 work *de Vita et Moribus Epicuri*, his eight books on Epicurean philosophy, and his 1649 commentary on the biography by Diogenes Laërtius, titled the *Syntagma philosophiae Epicuri*. It is likely Locke was also influenced by Gassendi, who in turn interested Condillac, especially his scientific approach to the history of ideas. I have a chain of knowledge of my own going but I don't tell d'Alembert. I am brought back from my musing when d'Alembert completes his answer to the question about the intentions of the 'men of letters' in relation to the origin and generation of ideas, 'Quite aside from the help this examination will give us for the encyclopaedic enumeration of the sciences and the arts, it cannot be out of place at the head of a work such as this.'[128]

D'Alembert is a confident young man in his mid-twenties. This is the first piece of extended writing he has done outside his chosen field of mathematics and natural science, although he was prone to write in a philosophical vein in 'introductions to his scientific treatises.'[129] It is a

---

[128] d'Alembert, J.L.R., *Preliminary Discourse to the Encyclopedia of Diderot*, p.5.

[129] Ibid., p.xxxix.

very impressive piece of writing. 'So how do you plan to deal with the question of knowledge and how we come by it?' He is into his answer before my last word has travelled to his ear 'We can divide all our knowledge into direct and reflective knowledge. We receive direct knowledge immediately, without any operation of our will; it is the knowledge which finds all the doors of our souls open, so to speak, and enters without resistance and without effort. The mind acquires reflective knowledge by making use of direct knowledge, unifying and combining it.'[130] 'When you are talking about receiving direct knowledge and using the analogy of entering through doors you mean through our senses.' 'Yes, all our direct knowledge can be reduced to what we receive through our senses; whence it follows that we owe all our ideas and sensations.'[131] 'It seems to me' I proceed 'that despite the number of times Bacon is mentioned in the *Encyclopédie*, including as appendices to your discourse, you owe a lot more to the thinking of Descartes.' He is on sure ground 'The fact of our existence is the first thing taught us by our sensations and, indeed, is inseparable from them. From this it follows that our first reflective ideas must be

---

[130] Ibid., p.6.

[131] Ibid.

concerned with ourselves, that is to say, must concern that thinking principle which constitutes our nature and which is in no way distinct from ourselves.'[132] In his carefully thought out reply he has reduced Descartes' *cogito ergo sum* to 'I have sensations therefore I am'. Unlike Descartes however, there is no God, an omission followers of Descartes soon notice with alarm. D'Alembert mentions a 'Superior Being' in passing but soon gets bogged down in trying explain the relation of mind to the external world, the world of objects. I ask if he can make this clearer 'The second thing taught us by our sensations is the existence of external objects, among which we must include our own bodies, since they are, so to speak, external to us even before we have defined the nature of the thinking principle within us. These innumerable external objects produce a powerful and continued effect upon us which binds us to them so forcefully that, after an instant when our reflective ideas turn our consciousness inward, we are forced outside again by the sensations that besiege us on all sides.'[133]

An image of a camera obscura keeps appearing in my thoughts as I am listening to d'Alembert. It is his

---

[132] Ibid., p.8.

[133] Ibid.

descriptions of the inside and outside and the attractiveness of the outside world, the instant recognition of the external world of objects that give us our bearings. It is like each instant presents the objective world to us in a new light, a tabula rasa. I have seen the illustration *Dessein – Chambré Obscure* in the *Encyclopédie* where two types of camera obscura are shown, plus an example drawing, but that doesn't quite explain why the metaphor's he has chosen to explain human existence resemble this visibility technology. It would be worth asking d'Alembert some questions around what he argues against when he presents his case for objective, direct knowledge, and where he got these ideas from. He has already indicated that Condillac has been closely involved in his writing, and it is no secret Condillac is inspired by Locke, and d'Alembert himself is familiar with the work of the English philosopher and political theorist, in particular his *Essay Concerning Human Understanding* of 1690 that was translated into French in 1700. Locke argues that all knowledge is derived from experience, either through the senses or on reflection of what is perceived by the senses. When Locke is setting up his argument 'Of Discerning, and other operations of the Mind' he uses the metaphor of a camera obscura to explain how it works,

I pretend not to teach, but to inquire; and therefore cannot but confess here again - that external and internal sensation are the only passages I can find of knowledge to the understanding. These alone, as far as I can discover, are the windows by which light is let into this dark room. For, methinks, the understanding is not much unlike a closet wholly shut from light, with only some little openings left, to let in external visible resemblances, or ideas of things without: would the pictures coming into such a dark room but stay there, and lie so orderly as to be found upon occasion, it would very much resemble the understanding of a man, in reference to all objects of sight, and the ideas of them.[134]

And he goes to the next stage of the usefulness of the metaphor in the section 'All ideas come from sensation or reflection',

Let us then suppose the mind to be, as we say, white paper, void of all characters, without any ideas:- How comes it to be furnished? Whence comes it by that vast store which the busy and boundless fancy of man has painted on it with an

---

[134] http://oregonstate.edu/instruct/phl302/texts/locke/locke1/Book2a.html#Chapter I

almost endless variety? Whence has it all the materials of reason and knowledge? To this I answer, in one word, from EXPERIENCE. In that all our knowledge is founded; and from that it ultimately derives itself. Our observation employed either, about external sensible objects, or about the internal operations of our minds perceived and reflected on by ourselves, is that which supplies our understandings with all the materials of thinking. These two are the fountains of knowledge, from whence all the ideas we have, or can naturally have, do spring.[135]

Thus Locke sets up his ideal space for the acquisition of knowledge, the images represented by the camera obscura, the process of storing knowledge in that prescribed space – the white paper - and finally the retrieval of it when required. In fact the writing of Locke abounds with metaphors drawn from visibility technologies including lenses and mirrors; so much so that I have drifted from my discussion with d'Alembert.

I tell him about my reading of Locke and ask if he directly uses passages from his work. He looks puzzled for a

---

[135] Ibid.

moment before saying 'of course I do' and then I remember the formalised rules of citation are not usual practice for a writer in the eighteenth century.[136] D'Alembert explains he uses a section of chapter one of *Essay Concerning Human Understanding* when he argues against innate ideas 'Why suppose that we have purely intellectual notions at the outset, if all we need do in order to form them is to reflect upon our sensations?'[137] 'Oh, and I talk about him directly' he continues 'in relation to Newton and metaphysics and how he didn't study books but rather contented himself with probing deeply into himself, so to speak, for a long time, he did nothing more in his treatise than to present mankind with the mirror in which he had looked at himself.'[138] We spend some time discussing the visibility metaphors and the camera obscura and I tell him that the encyclopaedic order they have created by relating the different kinds of knowledge to the various faculties of the soul and that the process of gathering, storing and disseminating

---

[136] In fact assigning authorship in the *Encyclopédie* is a difficult prospect. Many articles have no author and those that are marked are not always correct, while later printings left authorship out entirely. For a fuller explanation see *The Encyclopédie of Diderot and D'Alembert: Selected Articles*, Edited by J. Lough, Cambridge University Press, 1954, p.xiii.

[137] d'Alembert, J.L.R., *Preliminary Discourse to the Encyclopedia of Diderot*, p.7.

[138] Ibid., p.84.

science, replicates, to me, both their ideas on thought and the operations of a camera obscura. 'As my friend Diderot says in the section *Detailed System of Human Knowledge* physical beings act on the senses. The impressions of these beings stimulate perceptions of them in understanding,'[139] he relates before pausing, and then continues 'the sciences are the work of the reflection and of the natural light of men. Chancellor Bacon was therefore justified in saying in his admirable work *De dignitate et augmento scientiarum* that the history of the world without the history of scholars is the statue of Polyphemus with his eye torn out.'[140] In Greek mythology Polyphemus, the son of Poseidon, is a Cyclops, a one-eyed giant. D'Alembert's explanation sets me thinking about lenses, in particular the scioptric ball of the camera obscura and the 'ox-eye' experiment that Kepler and Descartes talk about.

After taking my leave from d'Alembert, I pause for an espresso and begin flicking through the marvellous

---

[139] Ibid., p.143. Schwab points out that the observations on Bacon and the outline of his system of knowledge were marked by Diderot's customary asterisk. See p.xxx of the introduction. By contrast Russell has d'Alembert as author, Russell, T.M., *Architecture in the Encyclopédie of Diderot and D'Alembert*, Scolar Press, England, 1993, p.9. I take Schwab as authority.

[140] Ibid., p.146.

illustrations in the *Encyclopédie* and think, it would be worthwhile animating the figures from these illustrations and have them come to life, to see how they move within the prescribed spaces they inhabit. The illustrations are carefully ordered as are the people represented, with only the occasional random character, like the middle distance figure sprinting from the falling roof tiles in 'Architecture: Tile Work', or the woman gazing out the window in 'Architecture: Tile Layer'. I am sure the engravers had a sense of humour but the *Encyclopédie* is above all a serious project, and they have mostly restrained themselves in the 2,885 meticulously detailed illustrations. The *Encyclopédie* tableau is an intellectual picture composed under a single point of view, that of the human figure and its endeavours. The complete human body with organs intact and properly arranged determines the true composition of knowledge. The *Encyclopédie* project, although sometimes intellectually uneven, is not thrown together at random but operates at a level of desire for a balanced composition, the aim harmony. What is selected is carefully placed and will remain in the field of knowledge, what is not will be silenced or at best sidelined. For example, in the engraving depicting cheese-making, the artist reduced the shaggy and 'disgusting' beard of the Germanic

Anabaptist herdsman (who went unshaven for religious reasons) out of deference to the sensibilities of French readers. The depiction of antiseptic cleanliness was favoured over the 'dark-fermentings' that actually accompanied the process. Jean Desmarets, who investigated the dairy industry, was aware there was no correlation between cleanliness of manufacture and the quality of cheeses, nevertheless had a tidy mind and was shocked by the filthy state Auvergnat herder's kept their beasts and dairies.[141] Importantly, like the application of perspective to the pictorial plane, the human figure becomes the measure of all things for the tableau, while representations of this figure take a sanitised and idealized form.

---

[141] Plate 12 Cheese 1 and Plate 13 Cheese 2, *Vol.VI, 'Addition a L'Économie rustiqué'*, in Gillespie, C.C., (ed), *The Diderot Encyclopedia of Trades and Industry*, Dover Publications Inc., New York, 1959.

CHAPTER FIVE

The Balloon

Almost out of nowhere the first balloon flights take place.[142] On June 5 1783 in Annonay France, Jacques Étienne and Joseph Montgolfier, with their first public display, send a large smoke filled, meticulously painted sky blue and gold linen and paper bag into the air to a height of about a mile and a half. It managed to stay aloft for ten minutes, much to the delight and amazement of the assembled local nobility and townspeople. The popular story has it that the Montgolfier brothers, who are from a family of wealthy paper manufacturers, have observed the ash rising from a fire and begun experimenting with model balloons. About three months later the first passengers make the trip as a duck, a

---

[142] There is some evidence that the Chinese used what could be described as hot air balloons in the Shǔ Hàn period (221-263) when airborne lanterns were deployed for military signalling. The following account is drawn from Nahun, A., *Flying Machine*, Collins Publishers, Australia, 1990, pp.8-9, Stafford, B.M., *Voyage Into Substance*, The MIT Press, Cambridge Massachusetts and London England, 1984, Faujas de Saint-Fond, *Méthode Aisée De Faire La Machine Aérostatique, (Ballon Expériences 1783)*, Liege: Chez Lemarié, Imprimeur-Libraire deffons la Tour, Proch l'Hotel-de-Ville, M.DCC.LXXXIV (1784), 2 parties en 1 volume, and Gillespie, R., 'Ballooning in France and Britain, 1783-1786', *Isis*, Vol. 75, No.2, 1984, pp. 248-268.

rooster and a sheep ascend for a flight that lasts eight minutes. A curious choice with two flighted and one four legged earth bound farm animal. On August 27 Jacques Charles and M. Robert send their hydrogen filled varnished silk balloon aloft to a height of four thousand feet. It remains in view for about an hour before descending in a field fifteen miles from Paris. A crowd of astonished farmers, fearing the craft was either an evil spirit or the moon which has somehow broken loose and fallen to earth, have destroyed it with farm implements. Their actions have prompted King Louis XVI to announce the importance of the balloon for the advancement of science and that the populace has no reason for fear. On September 19 another Montgolfier balloon is released, this time in Versailles, under the auspices of the Academy of Sciences, Paris. A large crowd is gathered, held back from the launching platform by armed soldiers. The nobility and scientists have the front rows and the best view while the general populace crowd the main square in their anonymous thousands.

It is exactly a month later and the first humans have ventured aloft. The King's official historian Rozier has persuaded the monarch that he, rather than a condemned criminal, should make the first flight. For caution the balloon is tethered to the ground but has achieved an altitude of eighty feet in a flight lasting four and a half minutes.[143] On November 21 Rozier and Marquis d'Arlandes – who helped Rozier persuade the King for the October flight – take off on an untethered flight from Bois de Boulogne. An unknown source has supplied the following detailed report,

> The envelope consisted of cotton and paper coated in alum as a form of fire proofing. Cords sewn into the fabric carried a wicker gallery at the base. The aeronauts stood on opposite sides of the gallery to balance it and maintain lift by pitchforking straw through two openings into a brazier mounted in the neck of the balloon. Each had a sponge and a bucket of water to put out fires in the envelope. Soon the Marquis noticed that the envelope has burned through in several places. He called out to descend but Rozier pointed out that they were above the rooftops of Paris. They put out the fires

---

[143] Faujas de Saint-Fond, *Méthode Aisée De Faire La Machine Aérostatique, (Ballon Expériences 1783)* dates this event October 19, other sources say October 15. I take Saint-Fond as authority.

in the envelope and tested the overheated suspension cords.[144]

Rozier and d'Arlandes clear Paris and after twenty-five harrowing minutes land in parkland near the present day Place d'Italie. On December 1 Charles and one of the Robert brothers ascend in their hydrogen filled balloon, also clear Paris and land in the countryside some twenty-five miles away. Thankfully the King's message has preceded them and apparently these intrepid aeronauts take a few bottles of champagne with them to celebrate with the locals, also insurance against damage.

Developments are happening so fast that I am having difficulty getting to each new event. I have seen the first crossing of the English Channel by Blanchard and Jeffries, the flight where they jettisoned all their equipment and even their clothes in order to keep the craft aloft, but missed the next one when Rozier, in his experimental craft with a hot air and a hydrogen filled balloon tethered together, is killed when his craft explodes thirty minutes into flight over the channel. Soon all the major cities in Europe have ballooning events, as

---

[144] Taken from the balloon display Science Museum London, no source acknowledged.

well as making an appearance in the United States. It is difficult not to get caught up in the excitement but I need to step back a bit in order to understand the shift in visibility that has taken place. It is worth talking to some of the spectator's and participants. The French meteorologist Bertholon is enthusiastic and voluble in his praise and it doesn't take much to get him to launch 'If voyages are made to other climes and under other skies, if these circumnavigations are so strange, so interesting, so practical in pushing back the frontiers of science through the comparison of exotic objects which they reveal, is it possible for us to believe that those undertaken above our earth, and in the vast stretches of the atmosphere, could be any less instructive and capable of piquing curiosity? Yes, soon we shall witness bold aerial navigators, the Columbuses, the Vasco de Gamas, the Bougainvilles, the Cooks, the Pages, animated by a noble ardour, thrusting themselves into the plains of the air, and embarking under the auspices of physics and of the Montgolfiers on aerostatic voyages into regions that seem prohibited to man...'[145] 'You are convinced that balloon flight will help science develop?' I

[145] Bertholon, P., *Des avantages que la physique, et les arts qui en dependent, peuvent retirer des globes aérostatiques*, Montpellier: Jean Mortel Aisné, 1784, quoted in Stafford, B.M., *Voyage Into Substance*, pp.24-25.

ask. 'I have no doubt that the cause of science will be helped by the superior method of observation offered by balloon flight. In order to record accurately all that is observed in flight I urge the use of a camera obscura in order to transcribe what is seen. For example I had a ravishing vision where I was transported into cloudy regions where weather and wind conditions changed before my very eyes. Here water vapours freeze and coalesce into snow, the moon's corona and the sun's parhelion become distinct, and the rainbow appears at last as a perfect circle rather than an imperfect arc.'[146] With so much going on and rapidly changing it is difficult to imagine a camera obscura operator managing to capture the spirit of the *expérience* of balloon flight. [147]

Lunardi, the first aeronaut to set sail from English soil,[148] to the delight of the 150,000 strong crowd told me that 'the earth appeared like before, like an extensive plain, with the same variegated surface; but with the objects

---

[146] Stafford, B.M., *Voyage Into Substance*, p.486.

[147] Stafford says the word *expérience* used for balloon flight is synonymous with experiment. See Stafford, B.M., *Voyage Into Substance*, p.22.

[148] Tytler made the famous flight in his Grand Edinburgh Fire Balloon in the Scottish capital in 1784 thus claiming the first flight in Britain.

rather less distinguishable.'[149] And Rozier, before his early demise, reported that when the balloon first burst out of the clouds 'an immense terrain suddenly unfurled in which all objects were confounded', and struggling to find words that adequately describe his experience 'enjoying reality rather than illusion born of a lie' falls back on known description of the balloon as 'a lofty *batiment*, an enveloping building or shelter imperceptibly carried along by winds and protecting him from them.'[150] Like Descartes Rozier retreats to a position of comfort to view the world. In his earlier flight of eleven thousand, seven hundred and thirty two feet he exclaimed that 'in a minute I passed from winter to spring.'[151] Rouland believes that balloons 'offer man the opportunity to penetrate to the centre of the atmosphere, and to traverse immense distances in a very short time, while voyaging at the whim of the winds.'[152] Baldwin, who went aloft thanks to a generous Lunardi, tells me that 'air gives form to things' and goes on to describe the circular panorama

---

[149] Quoted in Ibid., p.149.

[150] Taken from J.F. Pilatre de Rozier, *Premier expérience de la montgolfiére constuite par l'ordre du roi*, second edition (Paris: De l'Imprimerie de Monsieur, 1784), pp.13-14, quoted in Stafford, B.M., *Voyage Into Substance*, p.355.

[151] Ibid., pp.404-405.

[152] Monge, L., Cassini and Bertholon, *Dictionaire de physique*, Vol.1, Paris: Hotel de Thou, 1793, p.223, quoted in Stafford, B.M., *Voyage Into Substance*, p.311.

made available by the elevated position offered by a balloon.[153] He has even executed a number of drawings including one that shows the earth, clouds and atmosphere all arranged in circular form and describes his drawing to me 'The spectator is...suspended above the centre of the view, looking down on the amphitheatre or white of the clouds, and seeing the city of Chester, as it appeared throu' the opening: which discovers the landscape below, limited by surrounding vapour, to something less than two miles in diameter. The breadth of the blue margin defines the apparent height of the spectator in the balloon (viz 4 miles) above the white floor of the clouds, as he hangs in the centre, and looks horizontally round, into the azure sky.'[154] The vision offered is all-encompassing and round.

What has initially been a metropolitan event has soon found its way into the provinces, and not just through the chance visit of an off-course aeronaute. Chambery, where Joseph de Maistre is a provincial magistrate, may

[153] Baldwin, T., *Airopaidia Containing the Narrative of a Balloon Excursion from Chester, the Eight of September 1785, Taken from Minutes Made during the Voyage*, Chester: J. Fletcher, 1786, quoted in Stafford, B.M., *Voyage Into Substance*, p.222.

[154] Illustration from 'Baldwin's 'Airopaidia: containing the narrative of a balloon excursion from Chester, the eighth of September, 1785' published in Chester in 1786. http://www.ingenious.org.uk/site.asp?s=S2&DCID=10419257

be a relatively small centre, but it is not remote from the latest fashions. Following on from and encouraged by the Montgolfiers' early successes is an enthusiastic army cadet, Xavier, de Maistre's brother. It is early January 1784 and Xavier and others have organised a project to launch Savoy's first balloon. Joseph has been sent to Geneva to consult with the celebrated physicist Saussure on the technical details and he has also written a prospectus to enlist subscribers. The first flight attempted on April 22 was not a success, however the flight of today (May 6) Xavier and Louis Brun, the technical adviser, have been aloft. The launch was witnessed by a large crowd, and after their success the young heroes have been feted at a gala dinner hosted by President Maistre and Mme Brun, which has been followed by a ball.[155] The event has been covered by the local press and will soon help Xavier on a writing career that will secure him the position of director at the scientific department of the St Petersburg admiralty. The event has disturbed Joseph's equilibrium. In a letter he has written to Marquis de Barol he explains the dilemma he is facing in choosing a career 'Other times I vainly exhort myself as well as I can reason, modesty and

---

[155] Lebrun, R.A., *Joseph de Maistre, An Intellectual Militant*, McGill-Queen's University Press, Kingston, 1988, p.38.

tranquillity: a certain force, a certain indefinable gas lifts me like a balloon, despite myself. I lose myself in the clouds with Monsieur empyrean. I would *do it*: I would my faith, I do not know too much what I would do.'[156] A few lines further on in his letter he refers to a 'diagonal' which has been interpreted by a number of commentators to mean the 1789 revolution. Whatever it refers to his dilemma is clearly evident as he surrenders his future to the vagaries of the elements. In his writings he takes a 'panoramic rather than a particular view of history, seeking, not to describe and relate the unique event, but to find behind the sweep of history some logic, some divine pattern, that would satisfy the selfsame urge for design and order felt so strongly by Enlightenment thinkers. History has substituted for nature, but the object remains the same.'[157] Balloon flight offers the panoramic view de Maistre desires, while the particular and local is flattened out and rushed past. The broad sweep is more attractive to him because it covers more ground and allows him the space for the formation of a general design. It is however rudderless.

---

[156] Lively, J., *The Works of Joseph de Maistre*, The Macmillan Company, London, 1965, p.79.

[157] Ibid., p.39.

Of course the news of these amazing ascents has travelled quickly, paving the way for professional show people who have begun earning a living by ascending from public gardens and charging an admission fee. Large spectator numbers have been attracted - on one occasion some 400,000 watched – and a carnival atmosphere has developed. Unhappily sometimes these crowds have become disturbed after paying the fee only to find the flight has been delayed or cancelled. This has led to balloonists setting sail with faulty equipment or partly inflated balloons. They do this in preference to facing a disappointed mob. They do however leave a powerful impression on spectators.[158] One of the results has been the coining of the term *Balloonania* to describe events. There has been a blossoming of aeronautical-balloon art that has found expression in paintings, prints, decorated ceramics, fans, snuff boxes, medals, tiles, ivory boxes, bellows and numerous other articles. It could be described as a craze with 'balloon hats, balloon sweets and balloon pantaloons, and any number of bloated products' finding their way onto the market.[159] With the public becoming accustomed to the exploits of

[158] Taken from the balloon display Science Museum London, no source acknowledged.

[159] Heilbron, J., *The Rise of Social Theory*, Polity Press, Cambridge, 1995, p.57.

the aeronauts the performances are becoming more theatrical, elaborate and dangerous. For example I have seen fireworks discharged from a balloon at considerable risk, parachute drops, ascents on horseback and giant balloons setting sail.[160]

The impact of these occasions on social consciousness is significant. Not only does the advent of balloon flight offer entertainment and advertise the success of science, its significance is felt at all levels of the emerging metropolitan population. Those present have been involved in some of the first large gatherings of an urban mass to witness an event of spectacle. While the scientists demonstrate the importance of science over superstition (and religion), they rise above an assembled crowd whose individual features flatten into anonymity, and become transported into the ethereal elements. On the ground the assembled population shares in the excitement and risk of the *expérience*, while replicas offer transportable mementoes. These early 'argonaute aeriens', as they have been called, have become in an instant both observed and observer, the subject and object of an event of spectacle. The crowd watches the

[160] Taken from the balloon display Science Museum London, no source acknowledged.

aeronauts ascend into the elements while the aeronauts see a mass of people whose individual features become indistinguishable the further they rise. The moment presents both the increased all-round vision that science has been calling for and an obscuring of vision.

Given the nature of the experience offered the scientists by balloon flight, explanations of the panorama become fanciful and reflect more the dreams of the particular scientist rather than what is on view. Many of the descriptions of balloon flight are packed full of metaphors about abandonment to the feminine elements of nature, for example the etching depicting the flight of Charles and Robert over the Tuileries on October 1 1783 bears the caption 'The Moment of Universal Exhilaration,' with balloons offering a spiritual and sexual moment of ecstasy. Balloon flight is rudderless and fast, the landscape flattened, less distinct and transitory. This macroscopic view has important implications for the mapping of the terrain, both for topographical applications (aerial perspective became the supreme mapping implement as an accurate grid was applied to the world) and for ethnographic studies. With the surfaces to be mapped flattened into a plain, explanations are governed by a discourse of seeing from a distance. While such a perspective is exciting and easy – a mountain is difficult to climb and a telescope offers a side or partial view – and is believed to be both quicker and more accurate to apply, detail and difference are sacrificed in the interests of an overview. Individual features become indistinguishable from the surrounding mass.

CHAPTER SIX

The Lithograph

Daumier thoughtfully prepares another limestone block for his illustration in today's edition of Charivari. The stone is a special fine-grained variety cut from the Solnhofen limestone beds in Bavaria. After cleaning the stone he draws directly onto it with a black, oil-based crayon. His lines are direct and accurate and he makes good use of light and shade in order to give the drawing depth. It is a drawing of two astronomers in animated conversation in front of a telescope. The night sky is dotted with stars and the two figures are dramatically front-lit. I notice that the image is reversed, only because he signs his initials back to front. He carefully burns the image on the surface with acid then applies gum arabic to the smooth surface. Gum arabic is water soluble so only adheres to the non-oily sections, thus sealing it. It is now time for printing and Daumier steps back to allow the highly skilled craftsman to produce the lithograph. The relationship of trust between these two men is essential for a successful outcome. The stone is continually whetted with water which adheres to the gum arabic sections but avoids the oily drawing. Amazingly the oil based ink does the opposite as it is rolled over the surface until the acid etched segments are taken up with ink. The stone is then run through a firm, even press and the ink transfers from the stone to the paper. This process is being repeated a number of times which allows me to talk with Daumier about his theme for today. The caption prepared for the illustration has the headline A SATISFIED ASTRONOMER and the text follows 'Yes, my dear friend, I happened to discover a comet...and according to my calculations, I have every reason to hope that it will hit our planet in 45 days...'

'Is this drawing part of a series?' I ask. 'Yes' he replies 'we are nearly half-way through a series called *les beaux jours de la vie* (the best days in life). This particular print will appear on page three of today's Charivari.'[161] 'What is the series about?' 'It is designed to give the public insight into the calamities and misfortunes of the bourgeois as well as their little vanities. The railroads have triggered off a new trend called tourism, making it possible for Europeans and, to the horror of Parisians, also for people

---

[161] The following information is taken from The Daumier Register, a website dedicated to the works of Daumier, http://www.daumier-register.org

from the French provinces, to travel to Paris for sightseeing and shopping. At the same time, Parisians have the opportunity to travel to distant places in Brittany or Normandy. The development of many new hotels and restaurants are struggling to keep up and tourists have to deal with the uncertainty of getting a hotel room. The 'prix fixe dinner' has been a big success but due to the low quality of food, the guest often has to deal with unpleasant consequences after dinner. The imagination of the Parisian chefs knows no boundaries and meals made from cats, rabbits and other street game are sold as gourmet dinners. What I try to do is alert the readers to these little mishaps and tricks.'[162]

While we have been talking the lithograph and caption have been put together ready for publishing and the craftsman is cleaning the stone in preparation for the next drawing. 'And what about a *satisfied astronomer*?' I ask, while thinking that it seems a shame the drawing has been destroyed. Daumier, who is very matter of fact about the now blank stone, smiles, 'Comets have been much in the news and the first two pages of today's Charivari discuss reports of a recent meteorite storm in America and how just last year the great March comet

---

[162] http://www.daumier-register.org/werkview.php?key=1114

could be seen in daylight. The return of the comet Charles-Quint comet has been predicted for June 13, 1857 and it is expected that when it hits the earth it will be the end of the world. This topic provides fertile material for satire. And my drawing turned out quite well.' 'Do you write the captions?' He replies abruptly 'I sometimes contribute but really the caption is useless...if my drawing means nothing to you, then it must be bad; the caption will not make it any better. If my drawing is good, you will be perfectly able to understand it...so what is the use of the caption?'[163] I recover my poise to reply 'Of course I appreciate your art and drawing skill but when it is taken out of context the intent will be lost. I need the explanatory caption, and the background information you have given me adds another dimension'. He grunts assent but is already drawing on the next stone. It seems the process of drawing is the most important element to Daumier and his interest in the social and political debate lasts as long the drawing is on the stone. The same could be said of the satirical newspaper Charivari of course.

[163] Alexandre, A., *Daumier*, Laurens, 1888, cited in Picard, R., *Daumier and the University, Teachers and Students*, Boston Book and Art Publisher, Boston, Massachusetts, 1970, p.18.

I leave Daumier to his drawing and wander through his studio. Lithography has been available for about 40 years. Its invention has been attributed to the Austrian actor and playwright Senefelder who apparently responded to the financial problems he had with the printing of one of his plays by experimenting with this novel technique. It makes a good story but given the technical difficulty and skilled labour requirement it would more likely have driven Senefelder further into debt. By the 1820's most of the technical problems had been sorted out and lithography was embraced enthusiastically by artists like Delacroix, Géricault and Goya. Goya's bullfighting scenes have become well known in France and have inspired Géricault to make many lithographs of the poverty he saw on the streets of London on a recent visit. Delacroix has written in his diary 'Tried some lithography. Superb ideas for that subject. Caricatures in Goya's manner.'[164] His 1828 drawing *Faust* is perhaps the most outstanding of his works. Three things are going on here; the first is caricature, the second representations of contemporary social conditions, the third travel.

---

[164] Porzio, D., (gen.ed), *Lithography: 200 Years of Art, History and Technique*, The Wellfleet Press, New York, 1983, p.10. Goya executed these lithographs towards the end of his life when he was almost blind.

My reflection is interrupted by the sight of 36 poly-coloured terra cotta bust caricatures scattered across Daumier's work bench. They are about six to eight inches high. Two in particular catch my eye. The first is the deputy magistrate Lozére who is unmistakable given his dumpy figure and unshaven face. The same goes for Lefebvre the banker with his pointed head dominated by a huge nose. A bit further along, standing in the corner and covered by an old piece of cloth is the instantly recognizable character Ratapoil, his comic stance makes me laugh out loud. Daumier looks up from his drawing smiling, 'I see you have found my caricatures. I enjoy working in a three dimensional medium. Although drawing on stone has a direct feel and the use of light and shade gives depth, there is something more to holding a piece and moulding and turning it.' 'It is also easier to correct your mistakes' I reply. He laughs. 'Make sure you re-cover Ratapoil' he says 'Even though my friends love him I am afraid he may cause me trouble.'[165]

---

[165] Ratapoil appeared in Charivari and then as a clay sculpture. The character 'ratskin' was an agent-provocateur of Louis-Napoleon, a cudgel-carrying bully who was employed to stir up public unrest and convince people to return Louis-Napoleon to power, which occurred in 1852. Daumier was a Republican. The bronze copies we have today were taken from a later plaster model, the original lost.

Lithography enjoyed considerable success once the technical problems had been overcome 'Everybody produced lithographs, everybody bought lithographs, and the specialist firm of Lemercier in Paris became renowned for its lithographic reproductions. Lemercier's assistants were able to improve on even the clumsiest lithographic attempts brought in from all over the world.'[166] Included in these 'modest' works were caricatures and scenes from social life, illustrated books and albums, landscapes and genre scenes. Landscape scenes from France, Italy and North America, drawn from life, were reworked by the firm Lemercier. One of the most adventurous undertakings was Cailleux, Taylor and Nodier's *Voyages pittoresques et romantiques dans l'ancienne France*, a multi-volume series on French architecture sponsored by the French government. The authors were motivated by an awareness that historical French monuments, somewhat neglected because of the interest in Greek and Roman antiquity, needed to be preserved. French archaeology has its roots here. *Voyages pittoresques* produced some 2000 lithographs, taken from original drawings in pen and ink, pencil, wash, and watercolour,

[166] Porzio, D., (gen.ed), *Lithography: 200 Years of Art, History and Technique*, p.10.

between 1820 and 1878. One of the contributor's in the 1820's was Daguerre of whom we will hear more soon.

Genre scenes were another avenue explored with success by lithographers working in this period. The world of '*grisettes, lorettes,* and later of *cocottes* and *petit dames* provided inspiration for the thousands of former students who had become notaries or merchants in the provinces and were happy to rediscover in these prints the customs, attitudes, and witticisms of their youth. Boating parties, bathing on the Seine, and idyllic picnics in the countryside were other favourite subjects – themes that greet the arrival of Impressionism.'[167] Grisettes are young grey-clad milliners of easy virtue; lorettes are women of free agency, un-kept consorts and cocottes high staccato notes. These caricatures allow a typology of experience to emerge, a romantic rebuilding that replaces memory. Is it the experience memory or the lithograph that is being remembered and subsequently built on and embossed, replacing the original experience? Stendhal wrote of his recovered impressions of buildings triggered by etchings in his *Life of Henry Brulard* written in 1835-6. Re-presented realities mediated by time and typologies and

---

[167] Ibid.

taxonomies means direct access is unavailable. The purchasers of these genre lithographs believe they have recaptured times past. These particular lithographs awakened an interest in 'stories through pictures' that had been a feature of the 15th and 16th centuries. These were strip cartoons done on lithographic plates with a pen. Töpffer published his *Histoire de M.Jabot* in Geneva in 1883 and is said to have brought 'laughter to the whole of Europe.'[168] The 'nobility' of such expression was captured by Töpffer, where stories 'can be written by a succession of scenes portrayed graphically: this is literature in prints.'[169] The print or illustration had direct access to large numbers of people, even those with only rudimentary literary skills. Such easy access allowed the accompanying caption to do its work.

Genre lithographs incorporated official portraits, portraits of literary figures, portraits of women which also served as fashion plates, and also devotional images to be hung on the wall. Cheaply produced in large numbers this process ensured all classes in society could have these images in their homes. Whereas the *Encyclopédie* of Diderot and d'Alembert could only be afforded by the

[168] Ibid., p.12.

[169] Ibid.

privileged few, lithography, because it was not tied to the expensive and time-taking constraints of engraving, allowed the poor to obtain images of themselves and others. Not only did these images constitute reality for people, it also constituted their view of others through the same process. Although the early lithograph caricatures had been somewhat formal and cramped, as in the manner of its laboriously engraved copper antecedents, by the time of Daumier the medium has escaped such constraints and begun to express its own sensuous qualities. Under this influence, vast quantities of iconographical works, such as those adorning the walls of the poor, have entered the market. 'I am interested in the idea of caricature and how it works' I say to Daumier as he steps back from yet another completed drawing. 'So am I' he replies, 'but I have done enough drawing for today and am just heading out to meet Baudelaire and perhaps have a glass or two of absinthe. Baudelaire has quite a lot to say on the subject of caricature. Why don't you come along?'

As we arrive Baudelaire is showing Manet and Courbet a child's toy which he calls a Phenakistoscope, a mechanical optical contraption that imitates figures running. He is philosophising about children and toys 'All children talk to their toys; the toys become actors in the great drama of life, reduced in size by the *camera obscura* of their little brains. In their games children give evidence of their great capacity for abstraction and their high imaginative power.'[170] 'Ah Daumier!' he exclaims as we sit down with our first drinks for the evening 'one of the most important men, I will say not only in caricature, but in the whole of modern art. I want to speak about this man' and he looks directly at me 'who each morning keeps the populace of our city amused, a man who supplies the daily needs of public gaiety and provides its sustenance. The bourgeois, the business-man, the urchin and the housewife all laugh and pass on their way, as often as not – what base ingratitude! – without even glancing at his name.'[171] 'What is most useful about lithography?' I ask Baudelaire as soon as I find an opening. 'For the sketch of manners, the depiction of bourgeois life and the pageant of fashion, the technical means that is the most expeditious and the least costly...in trivial life, in the daily metamorphosis of external things, there is a rapidity of movement which calls for an equal speed of execution from the artist [...] lithography is admirably fitted for this

---

[170] Baudelaire, C., *The Painter of Modern Life and Other Essays*, A Da Capo Paperback, New York, 1986, p.198.

[171] Ibid., p.171.

enormous, though apparently so frivolous a task.'[172] 'Although you describe this task as frivolous you do promote the importance of the work Daumier and people like Gavarni are doing.' 'Yes' he replies 'we have some veritable monuments in this medium. The works of these two artists have been justly described as complements to the *Comédie Humaine*. I am satisfied that Balzac himself would not be averse from accepting this idea.'[173] There is general discussion about Balzac and his work and the fact that until recently he had been reduced to writing potboilers and serialized stories in order to keep food on the table. Daumier assured me Balzac would be here a little later for a coffee on the completion of his usual 15 hour writing day.

Baudelaire begins talking about Vernet as an entrée to his thoughts on the importance of caricature, '...for trivial prints, sketches of the crowd and the street, and caricatures, often constitute the most faithful mirror of life. Often, too, caricatures, like fashion-plates, become more caricatural the more old-fashioned they are [...] each pose and gesture has the accent of truth; each head

---

[172] Ibid., p.4.

[173] Ibid.

and physiognomy is endowed with an authentic style...'[174] he is into his stride now and begins describing the caricature types Grisettes and Lorettes, 'Gavarni created the *Lorette*. She existed, indeed, a little before his time, but he *completed* her.'[175] He continues describing the characteristics of a Lorette and finishes by saying Gavarni has had 'a considerable effect upon manners.'[176] 'You mean to say that people model themselves on these drawings?' 'Yes...not a few of those girls have perfected themselves by using her as a mirror, just as the youth of the Latin Quarter succumbed to the influence of his *Students*, and as many people force themselves into the likeness of fashion-plates.'[177] 'You make it sound like the process is only concerned with the frivolous and the shallow.' 'Not at all' he counters, 'The revolution of 1830, like all revolutions, occasioned a positive fever of caricature. For caricaturists, those were truly halcyon days. In that ruthless war against the government, and particularly against the king, men were all passion, all

---

[174] Ibid., p.166.

[175] Ibid., p.183.

[176] Ibid.

[177] Ibid.

fire.'[178] Daumier looks uncomfortable in his chair as Baudelaire waxes lyrical about Philipon's *La Caricature* of the early to mid-1830's, which depicted all the political elite. Daumier contributed a great deal to it, often under the pseudonym Rogelin, and spent six months in prison for his drawing *Gargantua*. As Daumier gets our next round of drinks Baudelaire reserves most praise for his depictions of the king who he calls the 'olympian and pyramidal *Pear*'. 'You will remember the time' Baudelaire continues 'when Philipon (who was perpetually at cross-purposes with His Majesty's justice) wanted to prove to the tribunal that nothing was more innocent than that prickly and provoking pear, and how, in the very presence of the court, he drew a series of sketches of which the first exactly reproduced the royal physiognomy, and each successive one, drawing further and further away from the primary image, approached ever closer to the fatal goal – the *pear*! 'There now,' he said: 'What connection can you see between this last sketch and the first?'[179] Daumier returns as Baudelaire, before waiting for the laughter to subside, continues 'Similar experiments were made with the head of Christ and that of Apollo, and I

---

[178] Ibid., pp.171-172.

[179] Ibid., p.172.

believe that it was even possible to refer back one of them to the likeness of a toad. But all of this proved absolutely nothing. An obliging analogy had discovered the symbol: from that time onwards the symbol was enough. With this kind of plastic slang, it was possible to say, and to make people understand, anything one wanted. And so that tyrannical and accursed pear became the focus for the whole pack of patriotic blood-hounds.'[180]

I am beginning to find it difficult to concentrate on what is being said. Baudelaire gets onto his favourite topic of Guys and is derided by Courbet who has become a lot more vocal. The tension between the two is evident, though they will remain friends for a lot longer than onlookers would imagine. The young Manet is quiet through most of this and Baudelaire almost ignores him. The ideas of de Maistre get a good hearing, with Baudelaire acknowledging his debt. The crucial role of images, the creation of character types and the forming of society through this process drifts in and out, with Daumier taking it all in with the eyes of a social recorder, as does Balzac when he finally arrives. The Goncourt's make an appearance and give me a good insight into the process of caricaturization and the work of Guys when

---

[180] Ibid.

talking about his working habits. After round six of the absinthe I silently take my leave, feeling like a blank block of limestone. Baudelaire's words chase me down the street 'Glorifier le cult des images (ma grande, mon unique, ma primitive passion).'[181] For me, the green fairy, rather than providing inspiration induces lassitude.

---

[181] Ibid., p.ix of the introduction.

CHAPTER SEVEN

The Diorama

In 1832 the German author and actor August Lewald attended a special breakfast in Paris behind Place du Château d'Eau (now Place de la Républic) in rue Sanson at the corner with rue des Marais. Being a keen diarist and letter writer he wrote of the occasion,

> We found ourselves under the eaves of a Swiss chalet. Below us we saw a small courtyard surrounded with buildings... On our left a goat bleated in his pen, and in the distance we heard the little bells of the herd ringing melodiously. But further away, what a view! The valley covered with snow, surrounded by gigantic mountains! There was no longer any doubt of the scene before us; I pointed out that before us lay Chamonix, 3,174 feet above sea-level...in the middle of the valley the majestic hump of Mont Blanc, 14,700 ft. high...Everyone was still standing filled with astonishment when another surprise succeeded

the first...We looked around and saw girls in peasant costume serving a country breakfast consisting of milk, cheese, black bread and sausage, while a man-servant poured out Madeira, port and champagne into crystal glasses. While at breakfast, we heard Alp-horns blowing a short solemn tune, after which a strong male voice down in the "valley" sang...in the dialect of the Chamonix valley. We were all greatly moved. "That is not painting---its magic does not go far as that!" exclaimed an English girl in the party. "Here is an extraordinary mixture of art and nature, producing the most astonishing effect, so that one cannot decide where nature ceases and art begins. That house is real, those trees are natural...Who is the artist who created all this?" "My friend Daguerre," I exclaimed enthusiastically. "His health!" We all clinked glasses. Daguerre approached and thanked us, obviously delighted at having been able to provide us with such a pleasant surprise in his Diorama.[182]

---

[182] Gernsheim, H & A., *L.J.M. Daguerre: The History of the Diorama and the Daguerreotype,* 2nd edition, rev. (New York: Dover Publications, 1968), pp.30-31.

Another conversation occurred at about the same time in Albany Street, Regents Park, London, between two spectators as they attended a similar event,

'I thought it was only to be a picture,' said Edward; 'but they must be real rocks and mountains made on purpose!' '*It is only a picture*,' replied Mr Finsbury. 'To convince yourself of this, you only have to change your position in the room, and you will see the objects are seen in exactly the same way, go where you will. For instance, there is a projection *represented* on a piece of rock in front. If a real projection, by moving across the apartment, you will see a little this way or that of the object behind it; but you find in vain that you rise from your seat, or stoop, or go from side to side; nothing different can be seen. It must, therefore, be a plain surface.'[183]

Edward is not convinced, but as Mr Finsbury appreciates, it is on the establishment of a single viewpoint that the diorama depends. It requires

---

[183] Bann, S., *The Clothing of Clio; A study of the representation of history in nineteenth-century Britain and France*, Cambridge University Press, Cambridge, 1984, p.27. This quote is taken from Taylor, Jefferys, *A month in London; or, Some of its modern wonders described*. London: Harvey and Darton, 1832.

considerable intellectual proof on the part of Mr Finsbury to reduce the illusion of reality to a mere crafted artifice. It is not a real town as Edward insists but rather a construction of paper, wood and colour, carefully illuminated from both the front and back. This large exhibition, about 40 x 72 feet, was a scaled up version Alberti's 1437 'enlarging and reducing machine,' which also appeared in della Porta's *Magica Naturalis* of 1558. Della Porta is said to have made a scaled up version himself and invited an audience inside for a viewing, a clever idea that, along with his other scientific and philosophical work, brought him to the attention of the Inquisition. Daguerre, who started out painting panoramas, made extensive use of the camera obscura when designing the scenery for his dioramas, enabling him to achieve both perspective and realistic detail.

A diorama (the word comes from the Greek 'a view through') is an exhibit or scene that displays modelled figures, buildings or objects that become smaller towards the back. Often accompanied by music and sounds, dioramas showing outdoor scenes have curved backdrops to give an illusion of depth. Lighting is a crucial element and can be supplied by strong reflected light which is hidden from the sight of spectators, or from partly transmitted light which offers glimpses through partitions, or through the use of semi-transparent panels. The overall aim of a diorama is to portray reality. Combining as it does careful illumination and arrangement of space, the diorama is a scenographic spectacle. As a device for organizing a particular effect, to represent reality, it takes as its subject matter ancient ruins, buildings, battles and so on drawn from history or eye witness accounts of events.[184]

Dioramas are part of the same family as panoramas and the curved backdrop for a diorama is panoramic painting. One of the earliest surviving panoramic works is a 1652 painting by Carel Fabritius titled *View of Delft, with a Musical Instrument Seller's Stall*. Apparently he used an elaborately ornamented surveying device called the 'Universal Instrument', made by Baldassare Lanci sometime around 1557, for this scene.[185] *View of Delft*

---

[184] See Thomas, S., 'Making Visible: The Diorama, the Double and the (Gothic) Subject' Robert Miles, ed., in *Romantic Circles, Praxis Series (Special Issue, 'Gothic Technologies: Visuality in the Romantic Era)* University of Maryland, 2005 and Walsh, P., 'The Painters and the Miracles: How Morse and Daguerre Created the Idea of the Media' http://web.mit.edu/comm-forum/mit5/papers/Walsh.pdf

[185] Bann, S., *The Clothing of Clio; A study of the representation of history in nineteenth-century Britain and France*, p.213. Lanci's surveying instrument may be operated as a theodolite, a geometric square, or a perspective machine. The perspective operation of the instrument was principally conceived for military purposes, whereby the surveyor's perspective drawing of the likes of a fortress was used to render the building's plan. For this operation, the instrument was originally equipped with two other accessories: a curved panel with a drawing sheet, and a

displays a sweeping curvature of roads and buildings, while objects close to the viewpoint of the artist and spectator are elongated. This particular view is taken from the corner of the Oude Langendijk and Oosteinde, looking roughly north west: in the centre is the Nieuwe Kerk and just to the left of it, in the distance, the town hall; both of these look much the same today. On the left, is the booth of a musical instrument vendor, with, in the foreground, a viola da gamba and a lute.[186] The viola looms towards us from the bottom right of the painting. Reminiscent of the effect a wide angled camera lens achieves it is possible Fabritius may have used a lens of some description. There is also speculation that this painting may have been part of a peep show or a perspective box. Measuring 15.4cm X 31.6cm, it would certainly fit in a box and would explain the distortions. On a larger scale Robert Barker designed a panoramic spectacle in 1787 that required an observer to step inside it to view and in 1789 he exhibited the first of his panoramas in London. It is worth remembering the word panorama didn't make its official appearance until about 1813 although Barker mentioned the word in a Times

---

sighting device with an ocular tube and a metal stylus. See http://brunelleschi.imss.fi.it/museum/esim.asp?c=500122.

[186] http://www.nationalgallery.org.uk

advertisement in 1792. Barker patented his invention in Edinburgh on June 19, 1787 as *La Nature à Coup d' Oeil* (Nature at a glance). The patent states,

> NOW KNOW YE, that by my invention, called La Nature à Coup d' Oeil, is intended, by drawing and painting, and a proper disposition of the whole, to perfect an entire view of any country or situation, as it appears to an observer turning quite round; to produce which effect, the painter or drawer must fix his station, and delineate correctly and connectedly every object which presents itself to his view as he turns round, concluding his drawing by a connection with where he began. He must observe the lights and shadows, how they fall, and perfect his piece to the best of his abilities.
>
> There must be a circular building or framing erected, on which this drawing or painting may be performed; or the same may be done on canvas, or other materials, and fixed or suspended on the same building or framing, to answer the purpose complete. It must be lighted entirely from the top, either by a glazed dome or otherwise, as the artist may think proper. There must be an inclosure within the said circular building or framing, which

shall prevent an observer going too near the drawing or painting, so as it may, from all parts it can be viewed, have its proper effect. This inclosure may represent a room, or platform, or any other situation, and may be any form thought most convenient, but the circular form is particularly recommended.

Of whatever extent this inside inclosure may be, there must be over it (supported from the bottom, or suspended from the top,) a shade or roof; which, in all directions, should project so far beyond this inclosure, as to prevent an observer seeing above the drawing or painting, when looking up; and there must be without this inclosure another interception, to represent a wall, paling, or other interception, as the natural objects represented, or fancy, may direct, so as effectually to prevent the observer from seeing below the bottom of the painting or drawing, by means of which interception nothing can be seen on the outer circle, but the drawing or painting intended to represent nature. The entrance to the inner inclosure must be from below a proper building or framing being erected for that purpose, so that no door or other interruption may disturb the circle on which the view is to be represented.

And there should be, below the painting or drawing, proper ventilators fixed, so as to render a current circulation of air through the whole; and the inner inclosure may be elevated, at the will of an artist, so as to make observers, on whatever situation he may wish they should imagine themselves, feel as if really on the very spot. In witness whereof, &c.[187]

The main emphasis is on constraining the observer and limiting the possibility of seeing the edges or border of the painting or drawing. The spectator must remain within the set framework for the device to work and there must be no visual interruptions or cues that allow a judgement on size or distance. The light is carefully controlled, mostly top lit, while leaving the observer in semi-darkness. To see the mechanics of the panorama is to break the illusion of reality. It must be said this patent is not a very good guide on how to build a panorama. Arrowsmith's 1824 diorama patent, complete with architectural drawings, is a lot more useful. Be that as it

[187] http://www.acmi.net.au/AIC/PANORAMA.html

may, in 1794 Barker's city-scape panoramas 'were joined in his new premises near Leicester Square by his stirring view of the *Grand Fleet at Spithead*, a scene which elicited an admiring response from Nelson himself.'[188] Another interested visitor to Barker's invention was Jeremy Bentham and there are obvious resemblances between his panoptic prison designs and these large circular panoramas.[189]

A few years before Barker, Philippe Jacques de Loutherbourg, who earned a living set painting and designing lighting for Garrick's theatre in Drury Lane London, had experimented with lighting and scene changes. He took his experiments one step further in 1781 and built his Eidophusikon, a moving panorama using three-dimensional sets, lighting, and sound effects to represent shipwrecks and natural wonders, such as Niagara Falls.[190] This display involved 'a system of moving pictures with a proscenium, which by a disposition of lights, coloured gauzes and the like,

[188] Kemp, M., *The Science of Art; Optical themes in western art from Brunelleschi to Seurat*, Yale University press, New Haven and London, 1990, p.214.

[189] Foucault, M., *Discipline and Punish*, Penguin Books, London, 1991 (1975), p.317, note 4.

[190] http://www.acmi.net.au/AIC/LOUTHERBOURG_BIO.html

created the impression of atmospheric effects at different times of the day – with appropriate musical accompaniments.'[191] A handbill from 1786 announces a miscellaneous exhibition of paintings in the Exhibition Rooms, over Exeter'Change, Strand,

> And in the evening will be presented the elegant and highly favoured SPECTACLE The EIDOPHUSIKON, invented and painted by Mr De Loutherbourg. In the course of which will be introduced the celebrated scene of The STORM & THE SHIPWRECK. The other scenes as usual. To conclude with the Grand Scene from Milton with the usual accompaniments. First seats 3s, second seats 2s. The doors will be opened at Seven and the Performance to begin at Half past seven.[192]

The tone is casual, the audience known and knowing. Five years of exhibiting and the spectacle is a standard feature of the evening life of London.

In April 1821 Daguerre formed a partnership with the landscape painter Charles Marie Bouton to build a 360

---

[191] Denvir, B., 'Controlled Sensibilities', *Art and Artists*, August 1977, pp.18-24, p.18.

[192] http://www.acmi.net.au/AIC/LOUTHERBOURG_BIO.html

seat diorama in Paris. Their aim was to produce a naturalistic illusion for the public, and make some money – the first aim was achieved, the second failed – with only insurance money from a timely fire saving Daguerre. Their first show opened on July 1822 in rue Sanson, the same place Lewald and his friends would later attend, with Daguerre providing scenes of *The Interior of Trinity Chapel* which were painted by Bouton, along with the Valley *of Unterwalden* and *Canterbury Cathedral* painted by Daguerre. These scenes, as well as showing great attention to detail and being paintings of worth (they took many months to complete), were depictions of places well known to the audience, as the following description attests,

> The visitors, after passing through a gloomy anteroom, were ushered into a circular chamber, apparently quite dark. One or two small shrouded lamps placed on the floor served dimly to light the way to a few descending steps and the voice of an invisible guide gave directions to walk forward. The eye soon became sufficiently accustomed to the darkness to distinguish the objects around and to perceive that there were several persons seated on benches opposite an open space resembling a large

window. Through the window was seen the interior of Canterbury Cathedral undergoing partial repair with the figures of two or three workmen resting from their labours. The pillars, the arches, the stone floor and steps, stained with damp, and the planks of wood strewn on the ground, all seemed to stand out in bold relief, so solidly as not to admit a doubt of their substantiality, whilst the floor extended to the distant pillars, temptingly inviting the tread of exploring footsteps. Few could be persuaded that what they saw was a mere painting on a flat surface. The impression was strengthened by perceiving the light and shadows change, as if clouds were passing over the sun, the rays of which occasionally shone through the painted windows, casting coloured shadows on the floor. Then shortly the lightness would disappear and the former gloom again obscure the objects that had been momentarily illumined. The illusion was rendered more perfect by the sensitive condition of the eye in the darkness of the surrounding chamber.[193]

---

[193] http://www.precinemahistory.net/1800.htm

Like the panorama, the diorama employs a careful arrangement of space, light and illusion, a central viewpoint and the restrained collaboration of the spectator. Like the stage, both the panorama and diorama require a careful placement of objects within a given space. Decisions are made about not only where to place particular objects so they can be seen in their best light, but also regarding the type and form of the objects to be included. Any object not carefully placed will not be seen at all or at best only partially. The objects selected need to be carefully sized to give the impression of depth and distance, as does the curved backdrop.

Frédéric Le Play, an early writer on the social, was the organizer of a number of Universal Exhibitions. These exhibitions were used to demonstrate the ascendant position of Western Civilization and offered a panoramic view analogous to the diorama. They used a fabrication of linear time, a vision of progress - requiring an overview of history, a central perspective convergence point and a central viewing point for the spectator.[194] Dioramas proved very popular at these exhibitions because they

---

[194] See de Cauter, L., 'The Panoramic Ecstasy: On World Exhibitions and the Disintegration of Experience', *Theory, Culture & Society*, SAGE, London, Newbury Park and New Delhi, Vol.10, November 1993, 1-23.

evoked reality, combined with scientific accuracy. Scenes from near and far were represented within the enclosed space of a well ordered and designed vision machine. One of the main task facing Le Play as he prepared for the exhibition of 1867 was how to house, order and display the variety and number of assembled objects. What was needed in this endeavour was a grid of the world. The exhibition of this particular year was housed under the roof of one building and incorporated a design similar to Bentham's panopticon. Surrounding the building was a park, which in later years was required for the display of machines and so on that spilled out of the main building.

Designed by Le Play and possibly Gustav Eiffel the building of 1867 had large concentric galleries arranged around an open courtyard. 'The space was divided', says a visitor, 'in concentric areas assigned to groups of similar products of all countries, and in radial sectors, each dedicated to a nation. So that one could, if one walked from the centre to the periphery through one of the sectors, pass all the products of the same country. But if one proceeded along a concentric gallery, one could see and study the same products of industry from different countries.'[195] An observer could stand in the central open space and gaze on the entire scene and then choose which section they wished to observe in more detail. They had available the ideal viewpoint, a view greatly enhanced by electric lighting which was used for the first time this year. This feature became a central attraction in later years. Vast numbers of people came to these exhibitions. Le Play invented a turnstile to count the number of visitors. In 1851 six million people attended and in 1900 some thirty-nine million were attracted.

I manage to have a few words with Le Play when I find him standing in the centre of the exhibition. The centre is a busy place as families and individuals gather here for an overview and in order to make decisions about where they will look first or next. It is the place to return for meeting up. It is also useful for Le Play because he can observe directly how well his design is working and what adjustments he will need to make for next time. 'What are the most important elements holding this arrangement together and what is your model based on?' I ask Le Play. He answers quickly and directly 'It is the family unit, held

[195] Faideau, F., 'Historiques des Expositions Universalles', in *Encyclopédie du Siecle*, Mongredin, Paris, 1900, p.19.

together by the moral values of thrift, discipline and *prévoyance*. The man, as head of the household is responsible for these values.' 'I understand thrift and discipline, but what is *prévoyance*?' I ask. 'It is the gift of foresight', and he looks at me doubtfully, 'this is not a socially learned ability but it can hopefully be learned through observation.' 'How can you know if someone has these qualities?' 'When *prévoyance* is combined with thrift and discipline the family budget will be balanced.' 'I then universalize these characteristics to include all of Europe.' 'Each social system is' he continues, 'charged with the responsibility for providing the mechanisms to make up for [any] deficiency.'[196] 'So at the centre of your enterprise you have the family unit, and outside that?' I ask, 'Directly overlooking the family are selected local élites who are guided by specially educated social scientists. Surrounding these people is the Christian religion with its important role in moral perfection and the most powerful means of achieving economic success.'[197] 'Arranged around religion at the centre of European civilisation are Germany, France and England. Other European countries are arranged at the periphery

---

[196] Le Play, F., *Les Ouvriers européens*, 2nd ed., (Tours, 1877-79), p.55.

[197] Silver, C.B., (Translator, ed.), *Frédéric Le Play on Family, Work and Social Change*, University of Chicago Press, Chicago and London, 1982, p.296.

of this main group.'[198] 'And how do you rank the nations?'
'The rank each nation occupies undoubtedly depends on
its material circumstances and the institutions that
govern it, but the elements essential to pre-eminence
belong to the moral order,'[199] he tells me. Clearly
demonstrated in his exhibition order and arrangement is
the privileged position of the European nations, followed
by those on the periphery, then follow countries from
further afield. Standing in the centre is the European
man with his wife and children standing dutifully behind
him. I turn to question Le Play about his idealized world
view, his recollections of social harmony he observed in
the Harz Mountains and in Luneberg, and how the
disordered reality I have seen outside the exhibition is not
represented here, but he has gone.

By the time of the 1900 Universal Exposition held in
Paris, the mechanics of the diorama was well advanced.
This particular exposition had several of these displays
on site. A Mareorama simulated a sea voyage from Nice
to Constantinople, proceeding via Venice. During the
performance two huge screens - 40ft high and 2500ft long

---

[198] Ibid., p.285.

[199] Ibid., p.296.

- were unrolled as the spectators stood on the deck of a pitching ship. The inventor of this system was the French Romanian painter Fred. Alexianu Hugo d'Alesi, best known for his exceptional travel posters, who apparently spent a year on board ship painting the various sections of the screens. A contemporary newspaper report goes to the heart of the device, 'Few visitors to the Exhibition will be able to resist the temptation...to make an inexpensive voyage which involves no hazards whatsoever, yet is so natural...even on the high seas, amid raging elements, one can get out and tread on terra firma at any moment.'[200] Other visual machines at the exposition were the Pleorama, Stereorama, Cineorama, the Lumiere Brothers photodrama and a simulated Trans-Siberian Railway. The railway was situated near the Russian and Chinese pavilions and was built by the Compagnie Internationale des Wagons-Lits, a Belgian company that deployed the first sleeping and dining car for long-distance train travel in Europe and, in 1883, initiated a service to Constantinople called the Orient Express.

---

[200] DeVries, L., Van Amstel, I., *Victorian Inventions*, John Murray, 1992, p.126.

CHAPTER EIGHT

Photography – some early stories

The pre-history (if I call it that reluctantly – progress the word appears in most texts) resembles soup, not a clear cut account of discovery, but different times and motivations and players, often disconnected in time and place and knowledge and endeavour from each other. However the same names keep appearing - which perhaps reflects my reading and sources. Porta appears with his camera obscura, Vitruvius noticed the sun altered and changed certain colours in his paintings so always placed them with a northern aspect, a disciple of Hermes, the alchemist Fabricius, threw sea salt into a solution of nitrite of silver and obtained a precipitate (chloride of silver) to which the alchemists of those times gave the name 'Luna cornea' or 'horn silver'. This white substance became instantly black when a ray of sunlight fell on it.[201] In 1777 the Swedish chemist Carl Scheele,

---

[201]Tissandier, G., *A History and Handbook of Photography*, Sampson Low, Marston, Searle & Rivington, London, 1878 (1973 Reprint), p.11.

discovered chloride of silver is more sensitive to blue and violet rays than to green and red, some nine years before he accidentally poisoned himself through his habit of tasting the chemicals he was experimenting with (a reasonably common practice at the time). Charles, from the balloon account, made use of a dark room in 1780 to produce rudimentary [photographs?] images on sheets of white paper which had been charged with a solution of chloride of silver. These images soon faded and turned to black when passed around during his demonstrations, though it has been reported he produced rough engravings that were placed on sensitised paper.[202] He also reproduced some engravings on sensitised paper, and in 1802 Thomas Wedgwood and Humphry Davy (who eventually died because of inhaling chemicals he worked with) published a treatise on the reproduction of objects by light called 'An account of a method of copying paintings upon glass and of making profiles by the agency of light upon nitrate of silver' delineating the profiles of woody fibres, leaves and the wings of insects. They were however unable to fix their pictures. Davy wrote 'All that is wanting is a means of preventing the lights of the picture from being afterwards coloured by

---

[202] Ibid., p.7.

daylight; if this result is arrived at, the process would become as useful as it is simple. Up to the present time it is necessary to keep the copy of the picture in the dark, and it can only be examined in the shade, and then but for a short time. I have tried in vain every possible method of preventing the uncoloured parts from being affected by the light. As for the images of the dark room, they were doubtless not sufficiently illuminated to enable me to obtain a visible picture with the nitrate of silver. It is that, nevertheless, which is the great point of interest in these experiments'.[203] Importantly though Wedgwood succeeded in obtaining impressions on leather (because of the accelerating influence of tannin or gallic acid) and on paper charged with a weak solution of nitrate of silver, and of crucial importance to events in the next fifty years, he carried his sensitive paper to the camera obscura. These images are not photographs and naming them so will not back-date the experiments or the results.

I return and ask Daniele Barbaro to demonstrate the camera obscura which he proposes be used for producing drawings in correct perspective. He says 'an old man's spectacle glass is suitable and a diaphragm may be

---

[203] Ibid., pp.12-13.

placed over the lens to restrict the aperture, and so increase the depth of field – although this reduces brightness.' He goes on to describe how 'Seeing, therefore, on the paper the outline of things, you can draw with a pencil all the perspective and the shading and colouring, according to nature, holding the paper tightly till you have finished the drawing.'[204] Seventeen years on another Venetian, Giovanni Benedetti, suggests a method for correcting the inversion of the image, by setting a plane mirror at 45 degrees to the direction of light coming from the lens.[205] By the late 17th century portable box-type camera obscura were modelled along these lines, the same arrangement of lenses and mirror's contained in a box as found in modern reflex cameras. And it is difficult not to be struck by the steady calmness in Vermeer's paintings of the mid-sixteen hundred's and to speculate on he and Leeuwenhoek talking and experimenting optics. They occupied the same time and space and place - along with Huygens - Delft.[206]

---

[204] Quoted in Steadman, P., *Vermeer's Camera*, Oxford University Press, Oxford, 2001, p.8.

[205] Giovanni Battista Benedetti, *Diversarum Speculationem Mathematicarum et Physicarum Liber*, Turin 1585, quoted in Ibid.

[206] Steadman, P., *Vermeer's Camera*, pp.44-58.

On 16 September 1824 Joseph Nicéphore Niépce wrote in a letter to his brother Claude; 'I have the satisfaction of being able to tell you that through an improvement in my process I have succeeded in obtaining a picture as good as I could wish...It was taken from your room at Le Gras with my biggest camera and my largest stone. The objects appear with astonishing sharpness and exactitude down to the smallest details and finest gradations. As the image is almost colourless, one can judge it only by holding it at an angle, and I can tell you the effect is downright magical.'[207] He also coined the term 'heliography' to identify the process by which he obtained these first images. The term is derived from the Greek words - *helios* meaning sun, and *graphein* denoting writing or drawing - encompassing both the source and the process of letting light record itself. These first images had short life-spans and soon faded. Perhaps they might have been called a heliotrope – escape by hiding or by becoming invisible (a power which the heliotrope, a precious stone, was believed to give to its bearer). Of course Niépce also invented a seat and handle bars for the bicycle, ha, how funny and how useful.

---

[207] http://www.hrc.utexas.edu/exhibitions/permanent/wfp/5.html

The optician Charles Chevalier has a shop on the Quai de L'Horloge in Paris. I decide to visit him because the shop is much frequented by amateurs and others, including the celebrated Daguerre. The year is 1825. A steady stream of people is coming through the door in order to purchase the latest dark room apparatus and to talk about the best lenses and who has discovered what. Amongst the well-presented appears a young man, poorly dressed, timid, and miserable and famished looking.[208] He approaches Chevalier and asks 'You are making a new camera in which the ordinary lens is replaced by a convergent meniscus glass: what is the price?' The optician's reply made his questioner turn yet paler. The cost of the object in question was doubtless far above his means as if it had been equal to the riches of Peru or California. He lowered his head sadly without speaking. 'May I enquire,' continued Chevalier, 'what you intend doing with a camera?' 'I have succeeded,' replied the unknown young man, 'in fixing the image of the camera on paper. But I have only a rough apparatus, a deal box furnished with an object-glass; by its aid I can obtain views from my window. I wished to procure your

_____

[208] The following discussion, slightly altered and abridged is in Tissandier, G., *A History and Handbook of Photography*. Tissandier quoted from Chevalier's article that appeared in *Guide du Photographe*, Paris, 1854.

improved camera lens in order to continue my experiments with a more powerful and certain apparatus.' Whilst listening to these words Chevalier said to himself, 'here is another of these fools who want to fix the image of the camera obscura!' he well knew that the problem engaged the minds of such men as Talbot and Daguerre, but non the less deemed it a Utopian dream. 'I know,' said he, 'several men of science who are engaged in this question, but as yet they have arrived at no result. Have you been more fortunate?' At these words the young man pulled out an old pocket-book which he was quite in keeping with his dress; he opened it a quietly drew out a paper which he placed on the counter. 'That,' said he, 'is what I can obtain.' Chevalier looked at it and could not control his astonishment; he saw on this paper a view of Paris as sharp as the image of the camera. It was not a drawing nor a painting: one might have said it was the shadow of the roof chimneys, and dome of the Pantheon. The inventor had *fixed* the view of Paris as seen from his window. Chevalier questioned the young man further, and the latter then drew from his pocket a vial containing a blackish fluid. 'You have here,' said he, 'the liquid with which I operate, and if you follow my instructions you will obtain like results.'[209] Downcast the young man left Chevalier's shop. Chevalier tried unsuccessfully to retrace the steps outlined by the young man but it is possible he left out a vital stage like preparing the sensitised paper in the dark, or some-such omission. He passed the information and the remaining liquid on to Daguerre, but he wasn't interested in either the information or the liquid, being more taken up with his own experiments. The young man was never seen again.

'Today – January 1839 – we announce' says Gaucherand 'an important discovery by our famous diorama painter, M. Daguerre. This discovery partakes of the prodigious. It upsets all scientific theories on light and optics, and it will revolutionize the art of drawing. M. Daguerre has found a way to fix the images which paint themselves within a camera obscura, so that these images are no longer transient reflections of objects, but their fixed and everlasting impress which, like a painting or engraving, can be taken away from the presence of objects. Imagine the faithfulness of nature's image reproduced in the camera and add to it the work of the sun's rays which fix this image, with all its range of high lights, shadows and half-tones, and you will have an idea of the beautiful

---

[209] Ibid., pp.34-36.

drawings which M. Daguerre, to our great interest, displayed.'[210] Daguerre, as always, is quick to step into the limelight recognising the chance to sell his invention via subscription, 'This discovery will,' he enthuses, 'give a new impulse to the arts...and far from damaging those who practice them, it will prove a great boon to them. The leisured class will find it a most attractive occupation...the little work it entails will greatly please the ladies...the daguerreotype is not merely an instrument which serves to draw nature...[it] gives her the power to reproduce herself.'[211] A crowd of interested onlookers soon gather around but Daguerre isn't having much luck selling his subscriptions. Perhaps it is the total price of 400,000 francs that has buyers balking, but he continues 'On January 15 I will have forty or fifty daguerreotypes placed on exhibition'.[212] Arago quietly comes up to Daguerre and whispers something in his ear and the two retire. Their conversation seems animated and it is well known Arago would prefer the French

---

[210] Quoted in Newhall, B., *Photography: Essays and images; illustrated readings in the history of photography*, Museum of Modern Art, New York, 1980, p.17, taken from *La Gazette de France* January 1839.

[211] Quoted in Scharf, A., *Art and Photography*, Penguin Books, England, 1968, p.25.

[212] Ibid. Surprisingly only 17 daguerreotypes made by Daguerre have survived. See Gernsheim, H & A, *A Concise History of Photography*, Thames and Hudson, London, 1971(1965), p.69.

government to award pensions to Daguerre and Niépce's son and place the invention in the public domain. As it transpires Daguerre doesn't hold his exhibition and the government do award lifetime pensions to the two men. I wonder if Daguerre actually has forty or fifty daguerreotypes to display, given the time to expose and process. He has *inside an antique shop* on display with its plaster reliefs of cherub's window-sill lit, and a few others, and *Notre Dame and the Ile de la Cite* but certainly not fifty. Almost forgotten in the uproar surrounding Daguerre is Hippolyte Bayard. He successfully produced natural images on paper (*dessins photogènes*) as early as February 5, 1839 and he holds an exhibition of some 30 photographs on June 24. Arago knows of this work and by some accounts has persuaded Bayard to postpone his announcement in order to favour his friend Daguerre. The French government pays Bayard 600 francs to purchase better equipment, and both they and Arago consign him to obscurity. Bayard doesn't go quietly. And Francis Bauer from the Royal Botanic Gardens, Kew, writes to the Literary Gazette to correct Arago's omission of Niépce's contribution when he addresses the Academie des Sciences and the Academie des Beaux Arts on August 19, 1839, and gently reminds them of Niépce's communication to the Royal Society in England in 1827.

Bauer includes examples of Niépce's heliographic process in his submission.[213] One of these is the *Set Table* of 1822, a reproduction of a glass negative plate coated with bitumen of Judeah. The table setting is for one.

Around 1840 the first portraits were being done. Antoine Claudet had bought Daguerre's rights to introduce photography to England and had discovered the properties of accelerating agents. This important development allowed Daguerreotypes of animate objects. Early photographs, because of the long exposure times required, could not capture movement. Streets full of people and carriages were emptied by the photograph to reveal only inanimate objects. There has been some speculation that these fleeting human images heralded impressionist painting. The word impressionism was being used for both the photographic process and its images prior to the painting movement.[214] For portrait sitters however the experience of having a photograph taken was akin to torture, the result falling well short of the pain they endured. Fifteen minute exposure times

[213] Bede, M., *Images: illusion and reality*, Australian Academy of Science, 1986, p. 35.

[214] Scharf, A., *Art and Photography*, p.350, notes. Many of the impressionist and realist school painters used photography, though few mentioned the fact.

meant a sitter - obliged to sit in the full sun - could not keep their eyes open nor could they smile. The invention of short-focus object-glasses reduced the sitting time to four or five minutes, still an uncomfortably long time. I decide to visit Lusset's photographic shop in the Place de la Bourse to watch one of these fast portraits being taken. Not without humour, Gaston Tissandier - who has accompanied me and describes the experience - calls the sitter a patient. 'The model took a graceful attitude, resting one hand on the back of a chair, and looking as amiable as possible. But the sun fell full in his eyes! The operator gives the final warning to keep perfectly still! The seconds pass, succeed each other, and seem to expand into centuries; the sitter in spite of all his efforts, is overpowered by the solar rays, his eyelids open and close, his face contracts, the immobility to which he is constrained becomes a torture. His features shrivel up, tears fall from his eyes, perspiration beads on his forehead, he pants for breath, his entire body shakes like that of an epileptic who wants to keep still, and the Daguerreotype plate represents the image of a poor wretch undergoing all the tortures of the ordeal by fire.'[215] 'Fortunately' says Tissandier, the discovery of

---

[215] Tissandier, G., *A History and Handbook of Photography*, pp.86-87.

accelerating substances has 'permitted Daguerreotype portraits to be taken with something of artistic feeling.'[216] Daumier has made over thirty five lithographs and drawings on the invention. Many of these deride the exposure time required to produce a portrait and show sitters constrained by various mechanical devices to keep them still. An obvious target is the closed eyes of the sitter that are reopened by careful retouching afterwards.

The mapping capabilities of the photograph have been applied immediately to record the local cityscape and foreign lands and the inhabitants of both. The painter William Müller is concentrating on detailed Egyptian studies of slave markets, shepherds, and intimate views of life in Cairo, while Emile Vernet and his pupil Goupil-Fesquet have just headed off to photograph scenes from the orient. It is November 1839, only a few months after Daguerre's announcement. Vernet reports he and his pupil are daguerreotyping 'like lions.'[217] The photographs they are taking are part of a commission from the opticians Lerebours who have requested Vernet and others to make daguerreotypes from Niagara to Moscow.

---

[216] Ibid., p.87.

[217] Scharf, A., *Art and Photography*, p.80.

More than 1200 views will be collected and published as *Excursions daguerriennes* in Paris in 1840-42.[218] This publication will be well received by artists and the general public. The photograph is also being used for artistic instruction because it demonstrates says Vernet 'correct aerial perspective by showing that the brightest parts of a view were in the foreground and not in the clouds or other parts of the sky as was thought the case by some artists.'[219] Baudelaire, for one, is not impressed with the attention to detail and precision of Vernet's paintings 'Who knows better than he the correct number of buttons on each uniform, or the anatomy of a gaiter or a boot. He has the memory of an almanac.[220] And Charles Blanc says 'Vernet's eye was like the lens of a camera, it had the same astonishing character, but also, like Daguerre's machine, it saw all, it reproduced all, without selection, without special emphasis. It recorded the detail just as the whole – what am I saying? – much better, because with Horace Vernet the detail always took on an exaggerated importance, so that inevitably it reaches a

[218] Ibid., p.82.

[219] Ibid.

[220] Ibid.

point where no trouble is taken to subordinate it, to give it its proper place and value.'[221]

Jean-Léon Gérôme has also travelled to Egypt to paint. He has recorded city scenes, landscapes and mildly erotic pictures of harems with rigorous objectivity and precision. As soon as he could Gérôme travelled with a camera and has both painted and taken photographs. Théophile Gautier praises these works in his writing of 1856,

> Photography, pushed today to the perfection that you know, relieves the artist from copying architectural and sculptural monuments, by producing prints of an absolute fidelity, to which the happy selection of the point of view and moment of time can give the greatest effect. Is that not also the direction in which Gérôme has taken his work; his powerful studies as a history painter, his talent as a draughtsman, fine, elegant, exact yet with lots of style, a special feeling which we would call ethnographic and which will become even more necessary to the artist in these days of

---

[221] Ibid.

universal and rapid travel when all people of the planet will be visited in whichever distant archipelago they may be hidden, all these things make Gérôme more suitable than any other to render that simple detail which up to now they have neglected, for landscape, monument and colour, modern explorations of the Orient – and man![222]

Photography, like all visibility inventions, unapologetically accompanied painting, drawing and lithographs from its inception. It soon found its way into the theatre of war – The Crimea 1855 - and became a reporting and propaganda tool.

I shuffle through the correspondence of William Henry Fox Talbot and find a letter he wrote to Arago on 29th January 1839,

Gentlemen,

In but a few days I shall have the honour of addressing to the Académie des Sciences, a formal

---

[222] Ibid., pp.82-83.

claim of precedence, of the invention announced by Mr Daguerre in its two principal points:

(1.) The fixing of the images of the *camera obscura*

(2.) The subsequent conservation of these images, in such a way as they might withstand direct sunlight.

Most occupied, at the moment, with a work on this subject, which will be read at the Royal Society the day after tomorrow, I shall content myself with asking you to accept the expression of all my consideration.

H.F.Talbot
Member of the Royal Society of London.[223]

A Fox Talbot letter to Francis Bauer two months later,

44 Queen Ann St London
Thursday 7th March

Sir,

---

[223] http://foxtalbot.dmu.ac.uk

Having observed in the Literary Gazette of last Saturday that you say, you have not yet seen any of my photogenic drawings, I will do myself the pleasure of calling on you on Saturday next, in company with my friend Professor Wheatstone, & will bring some of the specimens with me

I am, Sir, Your very obedient serv^t

H. Fox Talbot
Francis Bauer Esq,
F.R.S.Eglantine
CottageKew
H. F. Talbot Esqu Recd March 8th 1839[224]

Next to catch my eye is a letter he wrote to Giovanni Battista later that same year,

London 31 Sackville Street
21 August 1839

Dear Sir,

I am taking advantage of the occasion of the government sending a Courier to Italy to write you

---

[224] Ibid.

these few lines. As I fear that I will not be able to be present at the scientific meeting that will take place this year in Pisa, I would like to send you a package of my photogenic drawings as an expression of my esteem for the Italian scientists. Would you be willing to take the trouble of receiving this dispatch and of distributing it amongst the members of the scientific meeting as you see fit? If I receive your response in time I will be able to send the package at the end of September. Only tell me what is the best way to send it. If I am able to find some friend who is going to go to Pisa, I can entrust him with them.

This art of making images with light, which I have called Photogenic Drawing, was invented by me in the year 1834 – I have given a description of it in the Philosophical Magazine of March 1839. I showed them publicly for the first time on January 25th of this year.

It would be very agreeable for me, Sir, to renew our old acquaintance, which dates back to 1822, the period of my first journey to Italy – Will you not be coming to pay us a visit in England again?

Please accept my most sincere regards

H. F. Talbot
Professor Amici in Florence[225]

Part of the correspondence from Jean-Baptiste Biot to Fox Talbot on February 11 1840 is insightful,

> My aim in making this announcement was to preserve the complete independence of your rights over the representation of objects on sensitive papers, by keeping light and shade in their natural positions. There is a Mr Bayard here who works in finance who showed proofs made in this way in the camera obscura to much approval. This was after Daguerre's presentation but before his process was published. But Bayard has not made his method known; and since he wants, I believe, to use it more for his own profit than for the benefit of science, he has not ventured to make, or at least display, new proofs since the winter.[226]

---

225 Ibid.

226 Ibid.

The opening paragraph from a letter Friz Vogel wrote to Fox Talbot from Milan, Italy, 8 August 1851 gives us some idea of the significance of the invention,

Honourable Mr Talbot!

Thank you very much for your precious letter of <blank space>. Since I thought that the studio belonged to someone else, and since Mr Henneman does not have an idea of the form of my proposition, nor wish to know about it, and since it is true that language could be a great difficulty, even if polite society speaks French, I have taken your advice not to desire to come and work in London, and since it would involve a long journey and take up a lot of time, I do not wish to come to London to see the Exhibition either. I prefer to use the time to study and improve my art or rather your art since you are its inventor, and after the good Lord, we must thank you for the use of this beautiful art, which produces so much that is interesting and, by giving many men a livelihood, which also enriches their lives by nourishing the heart and the mind, as does true beauty.[227]

I reluctantly set Fox Talbot's correspondence aside and pick up Tissandier who left us with an exemplary record of the early years of photography and some of the exchanges and processes along the way. He is sure about the importance of the invention. The photographic image (whether still or moving) has dominated our view of the world since the mid nineteenth century in much the same way as the printing press allowed the printed word to dominate in the mid fifteenth century. I will sit quietly and listen to what Tissandier has to say in 1878.

> Hardly forty years have elapsed and the new invention has spread abroad and become so well known, that it has penetrated everywhere, in every civilised country, into the dwellings of the poor as well as of the rich. Unhappy indeed is he who cannot have recourse, for the picture of that which he loves, to photography, that sublime and beneficent art which gives us at such little cost the human visage in its exactitude, which presents to

---

[227] Ibid.

our eyes as in a mirror the scenery of distant lands, which lends its aid to all the sciences, which accompanies the astronomer into the depths of the heavens, the micrographer into the invisible world, and which even comes to the assistance of the besieged city, reducing its message to the easy burden of a bird![228]

The last line refers to the practice of using microscopic photographs, almost invisible to the naked eye, and attaching them to the tails of carrier pigeons, during the siege of Paris in 1870-71. The pigeons were transported by balloon from the capital. Careful attention needed to be paid to the quality of the magnifying lenses and the abilities of the operator was crucial. The messages were projected onto the wall by magic lantern and copied by clerks, ready for distribution.

An 1878 advertisement for The Ladies Camera, replete with etched illustrations, 'so called only from its Cleanliness, Convenience, and Portability' indicates how

---

[228] Tissandier, G., *A History and Handbook of Photography*, p.viii.

swiftly photography moved from a complex scientific process to an everyday tool. The manufacturers claim no previous knowledge of photography is required with this fifteen British pound apparatus,

> It has been designed especially to meet the want of *Amateurs, Professionals*, and, it is hoped, the fairer portion of society, who, would no doubt enjoy and practice the fascinating art, occasionally joining the sterner sex in their pleasant, healthful, outdoor excursional meetings, were they certain to be exempt from soiling their dainty fingers, carrying bulky apparatus, and last, but not least, could they hope to meet with the least amount of difficulty and trouble possible on their ambitious path. All these *desiderata* are, it is believed, fully obtained by the apparatus now brought out under the name of PHOTOGRAPHON.[229]

All aspects of the process, including introducing the chemicals, are mechanically controlled within the camera box simply by pulling a few cords. Regular lenses are

---

[229] Tissandier, G., *A History and Handbook of Photography*, un-numbered appendices.

included in the package for landscapes and portraits as well as special lenses for microscopic works.

BIBLIOGRAPHY

Alberti, L.B., *On Painting*, Penguin, England, 1972 (1436).

Alberti, L.B., *The Ten Books of Architecture*, Dover Publications, Inc. New York, 1986 (1775).

Baldwin, T., *Airopaidia Containing the Narrative of a Balloon Excursion from Chester, the Eight of September 1785, Taken from Minutes Made during the Voyage*, Chester: J. Fletcher, 1786.

Bann, S., *The Clothing of Clio; A study of the representation of history in nineteenth-century Britain and France*, Cambridge University Press, Cambridge, 1984.

Bardell, D., 'The First Record of Microscopic Observations', *BioScience*, Vol.33 No. 1, January 1983, pp. 36-38.

Baron, H., *The Crisis of the Early Italian Renaissance. Civic Humanism and Republican Liberty in an Age of Classicism and Tyranny* (in two volumes), Princeton University Press, Princeton, 1955.

Baron, H., *In Search of Florentine Humanism: Essays on the Transition from Medieval to Modern Thought*, Princeton University Press, Princeton, 1988.

Baudelaire, C., *The Painter of Modern Life and Other Essays*, A Da Capo Paperback, New York, 1986.

Bede, M., *Images: illusion and reality*, Australian Academy of Science, 1986.

Bertholon, P., *Des avantages que la physique, et les arts qui en dependent, peuvent retirer des globes aérostatiques*, Montpellier: Jean Mortel Aisné, 1784.

Chen-Morris, R., 'Shadows of Instruction: Optics and Classical Authorities in Kepler's "Somnium"', *Journal of the History of Ideas*, Vol. 66, No. 2. (Apr., 2005).

Christianson, J.R., *On Tycho's Island*, Cambridge University Press, 2003.

Contarino, R., (ed.), 'Canis' in *Apologhi ed elogi*, Genoa, 1984.

Crary, J., *Techniques of the Observer*, MIT Press, Cambridge, 1992.

Cronin, V., *The Florentine Renaissance*, The Folio Society London, 2001 (Collins 1967).

d'Alembert, J.L.R., *Preliminary Discourse to the Encyclopedia of Diderot*, The University of Chicago Press, Chicago and London, 1995(1751).

de Clercq, P., *At the Sign of the Oriental Lamp: The Musschenbroek Workshop in Leiden. 1660-1750*, Erasmus Publishing, Rotterdam, 1997.

de Cauter, L., 'The Panoramic Ecstasy: On World Exhibitions and the Disintegration of Experience', *Theory, Culture & Society*, SAGE, London, Newbury Park and New Delhi, Vol.10, November 1993, 1-23.

Denvir, B., 'Controlled Sensibilities', *Art and Artists*, August 1977, pp.18-24.

DeVries, L., Van Amstel, I., *Victorian Inventions*, John Murray, 1992.

Edgerton, S.Y., 'Alberti's Colour Theory: A Medieval Bottle without Renaissance Wine,' *Journal of the Warburg and Courtauld Institutes* Vol. 32 (1969), pp. 109-134.

Elkins, J., 'Response to Tomas Garcia Salgardo', *Leonardo*, Vol.29, No.1, 1996, pp.82-83.

Faideau, F., 'Historiques des Expositions Universalles', in *Encyclopédie du Siecle*, Mongredin, Paris, 1900, p.19.

Faujas de Saint-Fond, *Méthode Aisée De Faire La Machine Aérostatique, (Ballon Expériences 1783)*, Liege: Chez Lemarié, Imprimeur-Libraire deffons la Tour, Proch l'Hotel-de-Ville, M.DCC.LXXXIV (1784), 2 parties en 1 volume.

Ford, W., 'Development of our early knowledge concerning magnification', *Science*, Vol. 79, No. 2061, June 20 1934, pp. 578-581.

Ford, B.J., *The Revealing Lens: Mankind and the Microscope*, Harrap, London, 1973.

Foucault, M., *Discipline and Punish*, Penguin Books, London, 1991 (1975).

Gassendi, P., *The Life of Copernicus (1473-1543)*, Xulon Press, Fairfax, USA, 2002 (1647).

Gassendi, P., *The Mirrour of True Nobility & Gentility*, Infinity, Haverford, USA, 2003 (1641).

Gernsheim, H & A, *A Concise History of Photography*, Thames and Hudson, London, 1971(1965).

Gernsheim, H & A., *L.J.M. Daguerre: The History of the Diorama and the Daguerreotype,* 2nd edition, rev. (New York: Dover Publications, 1968).

Gillespie, C.C., (ed), *The Diderot Encyclopedia of Trades and Industry*, Dover Publications Inc., New York, 1959 (1751-1780).

Gillespie, R., 'Ballooning in France and Britain, 1783-1786', *Isis*, Vol. 75, No.2, 1984, pp. 248-268.

Gombrich, E.H., *Journal of the Warburg and Courtauld Institutes*, Vol.20, No 1-2, 1960.

Gould, S.J and Shearer, R.R, 'Drawing the Maxim from the Minim: The Unrecognized Source of Niceron's Influence Upon Duchamp', *tout-fait,* Issue 3, 2000.

Hacking, I., *Representing and Intervening*, Cambridge University Press, Cambridge, 1983.

Hainsworth, P., Lucchesi, V., Roaf, C., Robey, D., Woodhouse, J.R., 'The Languages of Literature in Renaissance Italy', reviewed in *Italica*, Vol. 67, No. 3 (Autumn, 1990), pp. 400-403.

Heilbron, J., *The Rise of Social Theory*, Polity Press, Cambridge, 1995.

I Libre della Famiglia, trs. Watkins, R.N. *The Family in Renaissance Florence*, Columbia (South Carolina), 1969.

Jefferys, T., *A month in London; or, Some of its modern wonders described.* London: Harvey and Darton, 1832.

Kemp, M., *The Science of Art; Optical themes in western art from Brunelleschi to Seurat*, Yale University press, New Haven and London, 1990.

Kepler's Somnium, The dream, or Posthumous Work on Lunar Astronomy, Translated, with a Commentary by Edward Rosen, Dover Publications, New York, 2003 (1967).

Lebrun, R.A., *Joseph de Maistre, An Intellectual Militant*, McGill-Queen's University Press, Kingston, 1988.

Le Play, F., *Les Ouvriers européens*, 2nd ed., (Tours, 1877-79).

Lively, J., *The Works of Joseph de Maistre*, The Macmillan Company, London, 1965.

Monge, L., Cassini and Bertholon, *Dictionaire de physique*, Vol.1, Paris: Hotel de Thou, 1793.

Munman, R., 'Optical Corrections in the Sculpture of Donatello', *Transactions of the American Philosophical Association*, Vol.75. Part 2, 1985.

Nahun, A., *Flying Machine*, Collins Publishers, Australia, 1990.

Newhall, B., *Photography: Essays and images; illustrated readings in the history of photography*, Museum of Modern Art, New York, 1980.

Onians, J., 'Alberti and Filareth', *Journal of the Warburg and Courtauld Institutes*, Vol.34 (1971), pp.96-114.

Philosophical Transactions, 1668, 1673, 1678 available online at JSTOR.

Picard, R., *Daumier and the University, Teachers and Students*, Boston Book and Art Publisher, Boston, Massachusetts, 1970.

Pilatre de Rozier, J, F., *Premier expérience de la montgolfiére constuite par l'ordre du roi*, second edition (Paris: De l'Imprimerie de Monsieur), 1784.

Porzio, D., (gen.ed), *Lithography: 200 Years of Art, History and Technique*, The Wellfleet Press, New York, 1983.

Rosen, E. 'Calvin's Attitude towards Copernicus' in *Journal of the History of Ideas*, New York, 1960, pp.431-441.

Russell, B., *A History of Western Philosophy*, George Allen and Unwin Ltd., London, 1946.

Russell, T.M., *Architecture in the Encyclopédie of Diderot and D'Alembert*, Scolar Press, England, 1993.

Scharf, A., *Art and Photography*, Penguin Books, England, 1968.

Silver, C.B., (Translator, ed.), *Frédéric Le Play on Family, Work and Social Change*, University of Chicago Press, Chicago and London, 1982.

Smith, M.A., 'What is the History of Medieval Optics Really About?'
La perspective curieuse, ou magie artificielle des effets merveilleux, November 5, 1651 F. Hilarion De Coste and Frere Ambroise Granjon publication.

Stafford, B.M., *Voyage Into Substance*, The MIT Press, Cambridge Massachusetts and London England, 1984.

Steadman, P., *Vermeer's Camera*, Oxford University Press, Oxford, 2001.

The Architectural Plates from the Encyclopédie, Diderot, D. (Ed.), Dover Publications, Inc. New York, 1995 (1751-1780).

The Encyclopedia of Diderot & d'Alembert Collaborative Translation Project, University of Michigan Library, online at http://name.umdl.umich.edu/

The Encyclopédie of Diderot and D'Alembert: Selected Articles, Edited by J. Lough, Cambridge University Press, 1954.

The Oxford Companion to Classical Literature, Howatson, M.C. (Ed), Oxford University Press, Oxford, 1989.

The Philosophical Writings of Descartes Vol.1, Cambridge University Press, Cambridge, 1985.

Tissandier, G., *A History and Handbook of Photography*, Sampson Low, Marston, Searle & Rivington, London, 1878 (1973 Reprint).

Thomas, S., 'Making Visible: The Diorama, the Double and the (Gothic) Subject' Robert Miles, ed., in *Romantic Circles, Praxis Series* (Special Issue, 'Gothic Technologies: Visuality in the Romantic Era) University of Maryland, 2005.

Vasari, G., *Lives of the Artists*, The Folio Society, London, 1993(1965 – first published 1550 and enlarged 1568).

Vitruvius, *The Ten Books on Architecture*, Trs. Morgan, M.H., Dover Publications, New York, 1960 (1914).

Walsh, P., 'The Painters and the Miracles: How Morse and Daguerre Created the Idea of the Media' http://web.mit.edu/comm-forum/mit5/papers/Walsh.pdf

Watkins, R., 'The Authorship of the Vita Anonyma of Leon Battista Alberti,' *Studies in the Renaissance* Vol. 4 (1957), pp. 101-112.

Wheelock, A.K., 'Constantijn Huygens and Early Attitudes to the Camera Obscura,' *History of Photography 1*, 1977.

www.ingramcontent.com/pod-product-compliance
Lightning Source LLC
Chambersburg PA
CBHW050119280326
41933CB00010B/1163